D0989460

VIRTUAL WORDS

Virtual Words

Language on the Edge of Science and Technology

Jonathon Keats

Property of
Baker College
of Allen Park

OXFORD
UNIVERSITY PRESS
2011

OXFORD
UNIVERSITY PRESS

Oxford University Press, Inc., publishes works that further
Oxford University's objective of excellence
in research, scholarship, and education.

Oxford   New York
Auckland   Cape Town   Dar es Salaam   Hong Kong   Karachi
Kuala Lumpur   Madrid   Melbourne   Mexico City   Nairobi
New Delhi   Shanghai   Taipei   Toronto

With offices in
Argentina   Austria   Brazil   Chile   Czech Republic   France   Greece
Guatemala   Hungary   Italy   Japan   Poland   Portugal   Singapore
South Korea   Switzerland   Thailand   Turkey   Ukraine   Vietnam

Copyright © 2011 by Jonathon Keats

Published by Oxford University Press, Inc.
198 Madison Avenue, New York, NY 10016

www.oup.com

Oxford is a registered trademark of Oxford University Press

All rights reserved. No part of this publication may be reproduced,
stored in a retrieval system, or transmitted, in any form or by any means,
electronic, mechanical, photocopying, recording, or otherwise,
without the prior permission of Oxford University Press.

Library of Congress Cataloging-in-Publication Data
Keats, Jonathon.
Virtual words : language on the edge of science
and technology / Jonathon Keats.
p.   cm.
Includes index.
ISBN 978-0-19-539854-0
1. Technology—Terminology.
2. English language—New words.
3. English language—Jargon.
4. Linguistic change. I. Title.
T11.K4184   2010
601'.4—dc22
2010010245

9 8 7 6 5 4 3 2 1

Printed in the United States of America
on acid-free paper

Contents

PREFACE

We live in an age of specialization. Perhaps in no sector are the constraints greater than within the sciences, where serious research in fields from molecular genetics to quantum mechanics depends on postdoctorate expertise. No longer is new knowledge casually acquired. There are no more gentlemen-scientists. There are no more natural philosophers.

I became a writer because my interests were too broad to abide specialized study. Writing is a general-purpose tool. Although English lacks the scientific rigor of tensor calculus and the technological vigor of assembly language, there is the advantage that prose communicates across disciplines. Writers are the envoys of modern society, and we are constrained only by the limits of our own curiosity.

At *Wired* magazine, where I write the monthly Jargon Watch column, my curiosity is focused on words. As an avowed generalist committed to surveying science and technology in the broadest sense, I cannot think of a more felicitous beat. The new language produced by a discipline provides a good indication of what that field considers fresh and important. (In this book examples of novel concepts include *microbiome* and *qubit*.) There's still more to be learned by examining how the language is formulated. (Again to cite examples from this book, consider the metaphoric motivations underlying *gene foundry*, or the case being made by *crowdsourcing*.) And given that language can impact thought, there's even a chance to observe the reciprocal influence of words on science and technology. (What will be the environmental consequences of naming our era the *anthropocene*? How have predictions of an impending technological singularity been shaped by how *singularity* is used in the description of black holes?) Looking at science and technology from the standpoint of language is not only opportune; it is essential to fully understanding the processes of discovery and invention.

Of course to be a writer is not necessarily to be a linguist or a lexicographer. In formal terms I am neither. Although both are worthy specializations benefiting from postdoctorate expertise, I insist on remaining a generalist, sampling multiple disciplines, writing from serial and cumulative observation. Therefore this book is not born of theory; a thesis has not been preconceived. These essays were undertaken in the spirit of dispatches from the field.

What emerges when these twenty-eight dispatches are surveyed together is a remarkable symbiosis between scientific and lexical innovation, a potent coevolution. Ideas inspire words, which inform ideas ad infinitum. In technology this kind of process is known as bootstrapping, and the word is apt, perhaps more appropriate than *coevolution,* because language is a technology, arguably our first, possibly our most resilient, developing with us as a species and facilitating all other technological advances from agriculture to the internet. The language of technology and science illuminates the science and technology of language. *Virtual Words* captures the internal workings of language from this technoscientific vantage.

The field I have chosen for this book overlaps with the ground I cover in *Wired,* though in most cases the words are different. For Jargon Watch each month I sort through a vast number of newspapers, magazines, and blogs, both mass-media and specialized publications, finding as many as one hundred notable words from which I select the four that seem most characteristic of the moment or most likely to have a future. In these pages my criteria are different. Novelty is not overwhelmingly important. (Some words, such as *steampunk* and *w00t*, have been around for decades, though they have recently entered the mainstream.) What mattered most in making this selection is that each word have a noteworthy history and that collectively they embody the diversity of scientific and technological language today.

You will find words that first appeared in scientific papers (*unparticle*) and words that gradually arose out of underground subcultures (*mashup*). You'll encounter words conjured to embody new concepts (*spime*) and words coined to suit new technological platforms (*k*). There are words invented to improve society (*copyleft*), to comment on it (*great firewall*), and for their own sake (*bacn*). The content of my essays reflects this variety. Historical context is the focus of my narrative on *copernicium*. *Conficker* is examined in terms of how and why the word was coined. My essay on *flog* considers why some terms catch on and others don't. *Panglish* provides an occasion to look more broadly at the future of English in our globally networked world.

This collection is far from exhaustive. For each one of these twenty-eight words there are countless others that could quite easily have taken its place. In the innovation category I might have included *cybrid* (a human-animal hybrid embryo). As an example of commentary I could have used *blackhawk* (a combative helicopter parent). For slang I was tempted by *sock puppet* (an illicit online alternate identity). These three terms might be fodder for another volume. If so, what I'll have to say about them will undoubtedly be different from what I would have written were they in the present collection. That's because the meanings of words in a book such as this one are too new to be settled. If these words endure, their significance will change (as is already happening in the case of *cloud* and *lifehacker,* for instance). Such is the power of language.

In time the significance of *Virtual Words* will also evolve. After the Twitterverse collapses, *tweet* will be a telling artifact of an early twenty-first-century preoccupation. If there is intelligent life elsewhere in the universe, and contact is ever made with aliens, *exopolitics* will serve more than conspiracy theories and idle speculation.

Like computer memory, science and technology have a tendency to overwrite their past. Theories are disproven. Machines break down. Words capture ideas in time, but meanings are fugitive. Here some representative terminology from a moment in history is fixed. Today is January 1, 2010. Or day zero for the lexical archaeologists.

Jonathon Keats
San Francisco, CA, and
Lago Maggiore, Italy

Acknowledgments

I am grateful to Mark Robinson and Chris Anderson at *Wired* magazine for entrusting me with Jargon Watch and to Elise Capron at the Sandra Dijkstra Literary Agency and Brian Hurley at Oxford University Press for helping to bring *Virtual Words* from conception to press. I'm also indebted to the Corporation of Yaddo in Saratoga Springs, NY, where portions of the manuscript were written. This book is dedicated to Silvia. *Tanto sempre.*

PART I

DISCOVERY

Examining a sliver of cork under a primitive microscope, the eminent seventeenth-century scientist Robert Hooke saw a structure of tiny compartments surrounded by thick walls that reminded him of the rooms inhabited by monks. He published his drawing of this surprising discovery, together with the first painstakingly accurate illustrations of flies, gnats, and fleas, in his 1665 book *Micrographia*. Those images were highly influential in his lifetime, revealing the future of biology to be in the microscopic details, yet it was his description of the cork that is best remembered: "These pores, or cells were not very deep," he wrote, "but consisted of a great many little Boxes, separated out of one continued long Pore." Cells still bear Hooke's monastic name today.

New discoveries are inherently strange. Established language, used by analogy or according to a familiar pattern, provides a bridge to the known world. The microbiome will not be entirely alien to someone who has studied the genome. Copernicium manifestly belongs to the same category as hydrogen and helium, even if its atomic weight cannot be deduced. Even unparticles reveal their characteristic foreignness by negation.

Yet patterns may be deceptive on the cutting edge of science, and analogies are often perilous. Atoms were allegedly atomic, indivisible, until J. J. Thomson discovered the electron. And as for the cells that Hooke described, they were indeed empty chambers, but that was because they were dead plant matter. It was the missing monks who were, metaphorically speaking, the basic units of life.

COPERNICIUM

The 112th element on the periodic table, named after Nicolas Copernicus.

The only accolade that American chemist Glen T. Seaborg cared for more than winning the Nobel Prize was having an element named in his honor. In 1994 his colleagues gave him that distinction, elevating the Nobel laureate to the status of helium and hydrogen. Over the next fifteen years, six more elements followed seaborgium onto the periodic table, bringing the total to 112. The last, enshrined in 2009, pays homage to Nicolas Copernicus.

Unlike Seaborg, Copernicus never sought such a tribute. Having already scored ample name recognition with the Copernican Revolution, he didn't really need it. If anything, by the time copernicium was recognized as an element, the periodic table needed him.

Copernicium is one of twenty elements containing more protons than the ninety-two naturally found in uranium. All twenty are made artificially in laboratories by colliding preexisting elements such as zinc and lead in a particle accelerator or

cyclotron. In some ten billion billion bombardments, two protons will fuse to make one atom of a new super-heavy element.
Typically the atom is unstable, lasting perhaps a millisecond
before decaying into lighter elements again. All of which makes
element fabrication a tricky enterprise, nearly as miraculous as
alchemy and considerably more contentious. Who synthesized
the first atom of an element, and therefore gets to name it?
Seaborg's UC Berkeley laboratory was the only one in the
business through the 1940s and 1950s, netting him ten elements,
including plutonium, for which he won the 1951 Nobel Prize
in Chemistry. By the 1960s, however, there was competition
from the Soviets, resulting in the so-called Transfermium Wars.
For several decades the periodic table became a political battle-
field rather than an intellectual commons.

Nothing could have been further from the table's Enlightenment origins. The product of empirical research and intended
to disseminate universal knowledge, a table of presumed
elements was first published by the French chemist Antoine
Lavoisier in 1789, arranging thirty-three substances, including
silver and sulfur and phosphorus, based on observed attributes
(such as "Oxydable and Acidifiable simple Metallic Bodies")
rather than according to philosophical precepts. This approach
reached its apex with the Russian chemist Dmitri Mendeleev's
1869 publication of a table that arranged the elements (by then
totaling sixty-five) in rows such that each column contained
substances with similar qualities. The periodic recurrence of
chemical attributes in atoms of progressively greater weight—
from which the periodic table derives its name—suggested that
some elements were still missing from the standard list. For
instance, there was a gap beneath aluminum. Rather than
changing his table Mendeleev predicted that an element would
be found. His apparent brashness was born of Enlightenment
diffidence: his table was not intended to impose order, but to

find order in nature, and if he discerned a natural pattern he was willing to give more credence to it than to received wisdom. Mendeleev's intuition was soon proven right. In 1875 the French chemist Paul-Emile Lecoq de Boisbaudran spectroscopically detected so-called eka-aluminum and promptly named it gallium to honor his nation (*Gallia* being Latin for France), while also slyly referencing himself (*gallus* being Latin for *coq*, French for rooster).

National pride, to say nothing of personal ego, became de rigueur over the next century with the discovery of elements such as germanium and polonium, named for Germany and Poland. Mendeleev's table initiated a gold rush—though of course gold had already been discovered, chemically speaking, and couldn't be claimed. The allure of new elements and the opportunity to name them led naturally to the literal transmutation of matter by Seaborg and company. By 1950 they'd made americanium and californium and berkelium, joined two years later by einsteinium and fermium, named for colleagues Albert Einstein and Enrico Fermi. Not that there was much boasting to be done at the time, since cold war paranoia about these radioactive metals required that they be kept secret even in name. Only as they were found to be useless, practically speaking, did their existence become known and the Transfermium Wars begin.

The Transfermium Wars were fought between UC Berkeley and the Joint Institute for Nuclear Research in Dubna, USSR, over who first fabricated the elements heavier than fermium, most viciously concerning elements 104 and 105, which were mired in mutual claims of scientific fraud. Though such heavy elements had no conceivable weapons potential, there was indisputable propaganda value, with particle physics becoming a nerdy equivalent to the space race. Accordingly the Americans called the two new elements rutherfordium and hahnium after physicists Ernest Rutherford and Otto Hahn, while the Soviets named the same substances kurchatovium and nielsbohrium after physicists Igor Kurchatov and Niels Bohr. Kurchatovium was by far the more incendiary of these, since Kurchatov headed the Soviet nuclear weapons program under Stalin. Nevertheless, *kurchatovium* appeared in Russian texts where *rutherfordium* was written in American books until 1994, when the International Union of Pure and Applied Chemistry (IUPEC) attempted to broker a compromise. The organization managed to infuriate everyone, including Glen T. Seaborg and the Institute for Heavy Ion Research in Darmstadt, Germany, which had independently synthesized every new element since 1981. The IUPEC proposed that element 104 be named dubnium after Dubna and rutherfordium be moved up to 106, displacing seaborgium on the grounds that Seaborg was still alive and therefore was ineligible to be an element. As for hahnium the IUPEC assigned it to number 108, which belonged to neither laboratory, since the Darmstadt group had fabricated it first and named it hassium to honor the Hesse region of Germany. For three more years the parties quarreled, finally agreeing that element 104 should be called rutherfordium, element 105 should be dubnium, and element 106 would remain seaborgium, with bohrium being bumped up to 107 and Otto Hahn, who was dead and therefore unable to complain, losing out entirely.

Ever since, the IUPEC has been working hard to regain respect. In 2002 the organization decreed, "In keeping with tradition, elements are named after a mythological concept or character (including an astronomical object); a mineral, or similar substance; a place or geographical region; a property of the element; or a [dead] scientist." Thus the arbitrary history of naming elements was codified in the finest bureaucratic fashion, and the trouble of wrangling with another Seaborg was conveniently eliminated.

Copernicium clearly fits the criteria and indeed was chosen specifically with the intention of restoring prestige to the periodic table. According to the Darmstadt team's leader, Sigurd Hofmann, the discoverers favored a non-German of indisputable international stature. (They also considered Galileo and several ancient Greek philosophers.) "We would like to honor an outstanding scientist, who changed our view of the world," Hofmann said in a press release issued by the lab, which cited the usual CV of Copernican achievements.

Yet it's for one of the astronomer's less famous accomplishments, understandably overlooked by the Darmstadt group, that the choice of Copernicus is most apposite. In 1526 Copernicus penned the influential *Treatise on Debasement*, considering a problem with money then circulating in Europe, which governments, eager to make more of it, coined in gold or silver of ever diminishing purity. Copernicus wisely perceived that the amount of gold or silver in a coin made no material difference, but that the quantity of money in circulation was critically important. Insisting on metallic purity, he claimed, was a way of curbing princes' greed, preserving the good faith necessary for a stable economy. Likewise it makes no material difference what we name the elements, but good science depends on the good faith of participants; in this sense nomenclature does matter, and few names in science have the intellectual purity of Copernicus.

Fabricating progressively heavier elements is more than a nationalistic pastime or a personal ego trip. Determining how many protons an atom can contain, and for how long, is a way of probing the nature of matter, the substance of everyone. The periodic table shows what we know of the world, but the names that we assign the elements reflect how we perceive ourselves.

MICROBIOME

The collective genome of all microbes hosted by the human body.

Sequencing the 3.2 billion base pairs of the human genome took thirteen years and cost $3 billion. Yet even before the genome was released in 2003, scientists were beginning to question whether an index of our DNA, however exhaustive, would genetically encapsulate *Homo sapiens*. One of the most articulate skeptics was Joshua Lederberg, a Nobel Prize–winning biologist who pioneered the field of bacterial genetics in the 1950s, while Watson and Crick were discovering the double helix. Lederberg publicly expressed his reservations in the year 2000, amid rampant hype about the Human Genome Project including a Clinton White House press conference. "Just as scientists study entire ecological systems to see how the various parts interact," he wrote in a syndicated editorial, "we must regard the human body as an extended genome. Its parts consist of the nuclear DNA genome (karyome), a chondriome (mitochondria), and what I call the microbiome: the menagerie of the body's attendant microbes. We must study the microbes that we carry within us and on our surfaces as part of a shared embodiment."

Lederberg coined the term *microbiome* to echo the older word *genome* and to establish a sort of equivalence.* If anything, his claim on behalf of single-cell organisms was understated. The average body is host to an estimated 100 trillion microbes, ten times the number of human cells. These microorganisms span some twenty-two phyla, with more than six hundred species living in the mouth and 150 species on each palm. Taken together that adds up to vast genetic diversity: human DNA contains approximately 23,000 genes. The microbiome may contain as many as 23 million.

Were these microbes mere fellow travelers the numbers would make no difference. (Because of their minuscule size microorganisms comprise less than 2 percent of our body mass.) What makes them significant from a human perspective—and makes their genetic legacy inseparable from ours—is that they perform essential functions in concert with human cells, facilitating digestion, for instance, and also aiding the immune system. We could not survive without our microbiome any more than our cells could subsist without mitochondria, metabolic organelles theorized to have originated as free-living bacteria billions of years ago. Challenging the hegemony of the human genome, the microbiome defies our definition of *human*.

In a paper published in the April 14, 2000, edition of *Science* Lederberg proposed a provocative alternative, suggesting that

---

* The first use of *genome* cited by the *Oxford English Dictionary* is in a German-language book by the botanist Hans Winkler, published in 1920. Since then *-ome* has become a favorite suffix for biologists, applied to just about anything that can be catalogued, especially when large-scale government funding is sought. Pre-microbiome coinages, of which there may have been more than a dozen, include proteome, transcriptome, physiome, and metabolome. Post-microbiome coinages are too numerous to count, but there are enough of them—from nutrigenomics and pharmacogenomics— to sustain the bimonthly publication of *OMICS: A Journal of Integrative Biology* and to garner 970 stimulus grants worth $625 million from the National Institutes of Health in fiscal year 2009 alone.

humans and microbes should together be construed as "a superorganism with the respective genomes yoked into a chimera of sorts." Four years later a headline in *Wired* magazine put it more bluntly: "People Are Human-Bacteria Hybrid." The *Wired* article was prompted by a paper published in *Nature Biotechnology* examining interactions between the human cellular apparatus and microbes and proposing that the microbial balance might influence everything from diabetes and heart disease to some cancers and neurological disorders. The authors, researchers at Imperial College, London, also suggested that the effectiveness of drugs might be aided or hindered by microbes, which mediate factors such as the gut pH. It was concerns such as these—not existential questions about what makes us human—that in late 2007 convinced the National Institutes of Health to launch a Human Microbiome Project to complement the Human Genome Project.

Announced just a couple months before Lederberg's death, the HMP was conceived as a five-year, $115 million project, touted by NIH Director Elias A. Zerhouni for its "potential to transform the ways we understand human health and prevent, diagnose and treat a wide range of conditions." Several simultaneous approaches were taken, including the complete sequencing of six hundred microbial genomes and the sampling of microbial communities living in prime habitats, such as the nose and the female urogenital tract. Almost immediately there emerged a microbial view of human anatomy distinct from our perspective, with bacterial populations radically differing over several inches of arm, yet being almost identical on the back of the knee and inside the navel. More surprisingly the microbial makeup of mouths on different continents was essentially constant, despite vast differences in diet. Taken together these data indicated a remarkable degree of human-microbe coevolution and reinforced one of Lederberg's motivating ideas.

"Those relatively few infectious agents that cause serious sickness or death are actually maladapted to their hosts," he wrote in *Science*. "Domesticating the host is the better long-term strategy for pathogens." Further NIH-funded research revealed an even closer symbiotic relationship. Microbiome species were observed to fight off microbial invaders that might potentially harm the host. And the absence of certain microbiome species—inadvertently eradicated by antibiotics or excessive hygiene—was correlated with human autoimmune disorders, including asthma and eczema.

These were important medical breakthroughs, earning laboratories another deserved round of NIH funding, but they also further narrowed the tenuous distinction between our microbiomes and ourselves. Here was evidence that our microbiome identifies with us closely enough that microbial self-defense includes protection of our whole organism, and also evidence that our immune system ceases to recognize us completely without our microbiome. We are a superorganism indeed, a bipedal ecosystem. Geneticists have responded by considering how evolution of the microbiome may supplement human evolution, which occurs far more slowly than societal changes, such as the development of agriculture and the industrial revolution. Given their ability to swap DNA by horizontal gene transfer, as well as their vast numbers and rapid reproductive cycles, bacteria evolve many orders of magnitude more quickly than their hosts. Their evolution may be the secret to their human hosts' physical adaptation to an unnaturally fast succession of external environments. In other words, our microbiome may allow us biologically to survive our cultural evolution. If so the microbes within us share responsibility for our development of farms and cities, the plow and the steam engine.

For practically as long as humans have practiced philosophy we have used reasoning as a tool to distinguish ourselves from everything else. For distinguishing features we have looked to our culture and intelligence, our evolution and genetics. Recognition of our microbiome undermines all of these conditions, since everything we are and everything we do is proportionally more microbial than animal. Accepting that our microbiome is a part of us, as our neurons and mitochondria are, alleviates one anxiety only to exacerbate another: unlike those neurons, or even the mitochondria, our microbiome isn't strictly delimited by our physical bodies. Our microbiome is promiscuous, passing between us with a handshake or a kiss and spreading into the environment all around us. Our biology confirms what philosophy has begrudgingly begun to accept: that our humanity is definitionally fuzzy, our attributes overlapping with those of other creatures, other forms of matter. Our microbiome belongs to our superorganism and makes us a part of an even larger superorganism: the planetary ecosystem.

Aside from a few new specialized words such as *microbiome*, our language has not yet come to reflect this worldview. For example, Helicobacter pylori is still known as a pathogen (from Greek words meaning "to produce disease") and vilified for its link to stomach cancer, despite the recent correlation of esophageal cancer with the *lack* of this bacterium in our gut. Nearly a decade before this complex relationship between microorganism and disease was revealed, Joshua Lederberg urged scientists to "drop the manichaean view of microbes." Yet the way we use language, the satisfaction we seek from words, is largely Manichaean, to the extent that we develop vocabulary to make distinctions. It takes the linguistic subversions of a poet to be meaningfully ambiguous, and poetry is a marginalized pursuit, whereas pathogenic rhetoric makes the evening news. For the superorganism *Homo sapiens* language evolves more gradually than scientific discovery.

Unparticle

A subatomic particle that by any conventional definition isn't one.

"All science is either physics or stamp collecting." So claimed Ernest Rutherford, the British physicist who discovered the atomic nucleus in 1910, touting the explanatory power of physics over the busywork of classifying elements or planets or animals. One hundred years later, the endless variety of matter postulated by physics—within the nucleus and throughout the universe—has far surpassed the inventories of the periodic table and solar system, leading particle physicists to refer to their domain as a bestiary and one textbook to be aptly titled *A Tour of the Subatomic Zoo.*

There are electrons and protons and neutrons, as well as quarks and positrons and neutrinos. There are also gluons and muons—the unexpected discovery of which, in 1936, led the physicist Isidor Rabi to quip, "Who ordered that?"—and potentially axions and saxions and saxinos. In this menagerie it's not easy for a new particle, especially a hypothetical one, to get attention. The unparticle, first proposed by American physicist

Howard Georgi in 2007, is therefore remarkable for garnering worldwide media attention and spurring more than a hundred scholarly papers, especially considering that there's no experimental evidence for it, nor is it called for mathematically by any prior theory.

What an unparticle is, exactly, remains vague. The strange form of matter first arose on paper when Georgi asked himself what properties a "scale-invariant" particle might have and how it might interact with the observable universe. Scale invariance is a quality of fractals, such as snowflakes and fern leaves, that makes them look essentially the same at any magnification. Georgi's analogous idea was to imagine particles that would interact with the same force regardless of the distance between them. What he found was that such particles would have no definite mass, which would, for example, exempt them from obeying special relativity. "It's very difficult to even find the words to describe what unparticles are," Georgi confessed to the magazine *New Scientist* in 2008, "because they are so unlike what we are familiar with." For those unprepared to follow his mathematics, the name evokes their essential foreignness.

The highly technical field of particle physics may seem a strange realm to find such linguistic whimsy. Yet the discipline's technicality encourages fanciful terminology precisely because it defies verbal description. The result is a vocabulary worthy of *Finnegans Wake*.

Consider the 1964 naming of the quark. Like the unparticle, the quark first came into the world as a mathematical rather than a physical entity. The physicists Murray Gell-Mann and George Zweig independently speculated that protons and neutrons might be construed as different combinations of a more fundamental form of matter coming in three varieties. Zweig called them *aces*. Gell-Mann preferred the sound of *kwork* or *quork*, which to his ear sounded like the noise made by a duck.

Exactly what theoretical physics had to do with ducks he never explained, but the weird sound he'd chosen was almost Joycean. In one of his "occasional perusals of *Finnegans Wake*" (as he relates in his memoirs) he happened upon the nonsense poem "Three quarks for Muster Mark." Noting the coincidence that the number of quarks in Joyce's poem matched the number in his own theory, he adopted that spelling, lending his particles a literary pedigree with which *aces* couldn't compete (even when it was later determined that quarks—Gell-Mann's, not Muster Mark's—came in more than three varieties).

The language of quarks has since expanded in a way that Joyce would likely approve. Quarks can have six different flavors: up, down, top, bottom, charm, and strange. And they come in three colors: red, blue, and green. Quantum chromodynamics accounts for how these colors interrelate, mandating that quarks be bound together in color-neutral combinations corresponding to how red, blue, and green light can be mixed to make white. The color names are metaphorical, quarks being imperceptible. Quark colors help us to grapple with what we cannot see, while flavor names such as *charm* remind us to take nothing for granted given quantum strangeness.

That strangeness easily trumps anything ever conjured in fiction. Alfred Jarry's absurdist nineteenth-century pataphysics (which he claimed was to metaphysics what metaphysics was to physics) seems downright reasonable compared to the Heisenberg Uncertainty Principle, and the sci-fi worlds of Stanisław Lem and Philip K. Dick have nothing on Schrödinger's Cat. It isn't for lack of writerly imagination. Descriptions of quantum phenomena leave written language in a tangle. The science fiction of the future will be written in mathematical equations.

In a sense that's what Howard Georgi has already done. Like a sci-fi writer, he's had the audacity to ask *What if . . . ?* and to let that be the premise for a fantasy that may yet (like so many sci-fi ideas of the past) turn out to be reality. The name that he's chosen, *unparticle*, captures this quality of play, which characterizes so much of quantum physics and differentiates it from the stamp-collecting branches of science.

Stars are numbered as they're discovered, and new elements are subject to strict standards of nomenclature. Even plants and animals, which have been named after presidents and rock stars, must be assigned Linnaean binomials, suitably Latinized. This makes sense because the purpose is to catalogue such entities in perpetuity. It isn't as clear as that in the case of quantum physics. Particles such as quarks are often named before their existence is known and may for years persist in this subatomic purgatory. The axion, for instance, was first postulated in 1977 and named after a popular cleaning detergent in the hope that the particle would "clean up" a problem pertaining to the strong nuclear force. Although still not detected in any experiment, and probably useless for its original theoretical purpose, the particle has suggested to researchers the possibility of saxions ("superpartners" of axions, according to supersymmetric theory, experimentally unverified in its own right) as well as axinos and even saxinos. To an outsider, quantum mechanics begins to resemble an advanced case of Tourette's. For an insider, though, the mutability of nomenclature is complementary to its informality. Unlike the names of elements, which sound almost sacred, the terminology of particle physics feels provisional, transitory.

That helps keep the field open to new possibilities. If unparticles exist, they may be the stuff of dark matter. Then their addition to the subatomic zoo might oust several other hypothetical particles. In fact most physicists would be thrilled to kill off the entire menagerie in favor of something more fundamental—perhaps strings or loops—as stunningly simple as the atoms proposed by Democritus in the fifth century BCE. This is the basic distinction between stamp collecting and physics that Ernest Rutherford insisted upon. The ambition of physics isn't additive but reductive. The paradoxical hope that more particles may lead to less is encoded in the language.

ANTHROPOCENE

The current geological epoch.

In geological time, the human life span is almost immeasurably brief. The seventeenth-century archbishop James Ussher famously calculated from biblical events that Earth was formed in 4004 BCE; scientists now estimate that the planet is 4.6 billion years old, and that the six millennia since the apocryphal Creation have probably contributed less than 10 millimeters of sediment to the geological record. Geological eras are unfathomable by ordinarily temporal measurements, such as the daily spin of the planet or its annual orbit, leading some scientists to adopt the galactic year—the 250 million terrestrial years it takes our solar system to rotate around the center of the galaxy—as a standard time unit. On that scale, *Homo sapiens* has been around for less than a week.

Yet as the technology to study the planet has improved, so too has the technology to alter it. Earth increasingly disproportionately bears our imprint, as if geological time were being accelerated to the beat of our biological clock, with the consequence that the planet seems increasingly mortal, its legacy and ours entangled. In geological terms we are in the Holocene epoch—a designation formulated from Greek roots meaning "wholly recent," officially adopted at the 1885 International Geological Congress—and have been in the Holocene for the past ten thousand years. The question, given all that we've done to the planet, is whether the label remains valid, or whether we've now buried the stratum of our Neolithic ancestors beneath our own rubbish.

The atmospheric chemist Paul Crutzen was the first to effectively challenge the conventional geological thinking. In a 2003 interview with *New Scientist* he recollected the circumstances: "This happened at a meeting three years ago. Someone said something about the Holocene, the geological era covering the period since the end of the last ice age. I suddenly thought this was wrong. In the past 200 years, humans have become a major geological force on the planet. So I said, no, we are not in the Holocene any more: we are in the Anthropocene. I just made up the word on the spur of the moment. But it seems to have stuck." Indeed, although *Anthropocene* lacks official status, the name is now so commonly used in scientific papers that many researchers don't feel compelled to explain what they mean by it.[*]

One reason it stuck, surely, is that Crutzen was a 1995 recipient of the Nobel Prize for his work on the man-made hole in the ozone layer, giving his voice authority in environmental matters. As important, though, is the careful support he subsequently provided for his spontaneous neologism. In a paper he

---

[*] Geologists have acknowledged in passing the effect of humans on their environment since at least the 1870s, when Antonio Stoppani wrote of the "anthropozoic era," defined by a "new telluric force, which in power and universality may be compared to the greater forces of earth." Other terms have since come and gone, most recently *Homogenocene* in the late 1990s, but *Anthropocene* is the first to enter the vernacular. Official endorsement falls under the jurisdiction of the International Union of Geological Sciences, which operates on a geological timescale and has not yet even taken the matter under consideration.

coauthored for the May 2000 International Geosphere-Biosphere Programme newsletter, he considered factors such as the vast accumulation of greenhouse gases in the atmosphere and the dramatic decrease of species diversity in rainforests. He also noted that 30 to 50 percent of the global landscape had been altered by human activity. "Without major catastrophes like an enormous volcanic eruption," his paper concluded, "mankind will remain a major geological force for many millennia, maybe millions of years, to come. To develop a world-wide accepted strategy leading to sustainability of ecosystems against human induced stresses will be one of the great future tasks of mankind." In other words, the term *Anthropocene* was meant to serve as a call to action, not just a physical observation.

This is highly unusual. Although scientists have periodically introduced politically loaded terms into public discourse—Crutzen himself coined *nuclear winter*—the role of science since the Enlightenment has been descriptive rather than prescriptive, and taxonomic terminology has generally reflected this distinction. The language of geology is typically functional. Names of epochs are often informative—at least to those who know Greek—as in the case of the Eocene, Oligocene, Miocene, Pliocene, Pleistocene, and Holocene, which provide highlights of speciation over the past 55 million years. (Eocene uses the Greek root for *dawn*, heralding the birth of modern mammals. Pliocene uses the Greek root for *more*, referring to the bumper crop of new mollusks.) Geological epochs are subdivisions of periods, which traditionally reference the places where characteristic rocks were first collected, as in the case of the Jurassic, named for the Jura Mountains, or the Cambrian, named for Cambria in the British Isles. And periods are subdivisions of eras, which tend to give basic chronological cues, as in the case of the Paleozoic, Mesozoic, and Cenozoic, referring to old, middle, and new life forms.

These three naming systems don't really fit together, nor are they truly systematic. There are abundant exceptions, such as the Cretaceous period, named after the chalk formed some hundred million years ago by single-celled coccolithophores, rather than for the Paris Basin where the chalk deposits were first discovered. Moreover some periods have had more than one name because independent discoveries of the same strata were made in more than one place. This etymological muddle is the freeform result of a three-centuries-old discipline encompassing a surface area of 510 million square kilometers over 4.6 billion years. Discoveries were often made in the field, and names were given based on what researchers deemed significant. Like the subject under study, the language of geology is layered, with older strata sometimes obducting newer, and a thorough etymological survey would provide a rich intellectual history.

Anthropocene lands atop this past, in changed circumstances. We have seen Earth from space and networked the planet with the world wide web. Modern geologists are by no means omniscient, but Earth science can now claim a truly global perspective, broadened by atmospheric measurements. We can monitor the rise of carbon dioxide levels due to fossil fuel consumption and observe the effect on ocean acidification, killing off corals, slowing the deposition of reef limestone on ocean floors. Reef limestone was characteristic of the Holocene, and the University of Leicester geologist Jan Zalasiewicz and colleagues have proposed that future planetary scientists will see its sudden decline as indicative of a new epoch.

Zalasiewicz and his collaborators have published the most comprehensive case to date for official adoption of *Anthropocene* in the February 2008 issue of the Geological Society of America's *GSA Today*. Their paper considers a whole host of potential markers, formally known as Global Standard Sections and Points (GSSPs), and colloquially called golden spikes. In addition to the decline of reef limestone, there's the increased erosion due to modern agriculture and, as coauthor Andy Gale bluntly put it to the BBC, "a hell of a lot of concrete." Each of these standards would presumably set future geologists' golden spike at a different section and point, and Zalasiewicz, like Crutzen, has been wary of choosing one.* In fact Zalasiewicz has instead suggested somewhat arbitrarily beginning the Anthropocene in the year 1800, when the world population reached one billion. As he argued in *GSA Today*, "This would allow (for the present and near future) simple and unambiguous correlation of the stratigraphical and historical records and give consistent utility and meaning to this as yet informal (but increasingly used) term."

Such a departure from geological convention is necessary because *Anthropocene*, unlike *Eocene* and even *Holocene*, anticipates how the future will view the present. For the first time we are classifying the geology of our own making; for the first time a species on this planet has the ability to do that.

If *Anthropocene* doesn't measure up to Enlightenment principles, it's because Enlightenment principles don't measure up to the Anthropocene. Objective observation is contingent on detachment from the system under observation. Our effect on the planet is too dramatic, and our scientific research too mired in the technology that alters it, for geology to harbor Enlightenment idealism. At a global level science is inherently political. By frankly acknowledging the autobiographical future of geology and insisting that we take responsibility, the term *Anthropocene* is a golden spike in the language of science.

---

* Geologists have argued that the Anthropocene should begin as far back as eight thousand years ago, marked by the clearing of forests to grow crops, and as recently as the 1960s, marked by the radioisotope signature of the hydrogen bomb. There is no golden standard for golden spikes, which have referenced fossils from mass-extinction events and even the anomalous level of iridium caused by a meteor strike.

PART II

INNOVATION

Developing a process for making pictures by projecting light onto chemically sensitized paper, William Henry Fox Talbot dubbed his invention *photogenic drawing*. His collaborator, John Herschel, preferred a crisper designation. In a paper he read before the Royal Society on March 14, 1839, discussing Talbot's experiments and his own, he made the first public reference to *photography*.

The word sounded scientific, more fitting for a new technology than Talbot's antiquated artistic phrasing. So thoroughly did *photography* embody the age, and so completely did photography mark the era, that just nine years after Talbot's invention was announced, Thackeray's *Vanity Fair* was being called "a literal photograph of the manners and habits of the nineteenth century."

So what are we to make of the names given to innovations in our own epoch? Terms such as *gene foundry* and *in vitro meat* evoke the past rather than the future, and *cloud* is effectively ancient. This may be a sign that our technology has become so advanced, and so abstract, that we need antique metaphors to talk about it.

Or it may simply be shrewd marketing. Shortly after Talbot worked out his process of printing with silver salts, Herschel developed a similar method for making photographs using bright blue potassium ferricynide. He called them *cyanotypes*. Yet only after his death did they meet success as a technology (and also a metaphor) when they were put to work in the construction industry, rechristened *blueprints*.

The Cloud

The web, viewed as a public utility.

"It's really just complete gibberish," seethed Larry Ellison when asked about the cloud at a financial analysts' conference in September 2008. "When is this idiocy going to stop?" By March 2009 the Oracle CEO had answered his own question, in a manner of speaking: in an earnings call to investors, Ellison brazenly peddled Oracle's own forthcoming software as "cloud-computing ready."

Ellison's capitulation was inevitable. The cloud is ubiquitous, the catchiest online metaphor since Tim Berners-Lee proposed "a way to link and access information of various kinds" at the European Organization for Nuclear Research (CERN) in 1990 and dubbed his creation the World Wide Web. In fact while many specific definitions of cloud computing have been advanced by companies seeking to capitalize on the cloud's popularity—Dell even attempted to trademark the term, unsuccessfully—the cloud has most broadly come to stand for the web, a metaphor for a metaphor reminding us of how unfathomable our era's signal invention has become.

When Berners-Lee conceived the web his ideas were anything but cloudy. His inspiration was hypertext, developed by the computer pioneer Ted Nelson in the 1960s as a means of explicitly linking wide-ranging information in a nonhierarchical way. Nelson envisioned a "docuverse" which he described as "a unified environment available to everyone providing access to this whole space." In 1980 Berners-Lee implemented this idea in a rudimentary way with a program called Enquire, which he used to cross-reference the software in CERN's Proton Synchrotron control room.

Over the following decade, machines such as the Proton Synchrotron threatened to swamp CERN with scientific data. Looking forward to the Large Hadron Collider, physicists began voicing concern about how they'd ever process their experiments, let alone productively share results with colleagues. Berners-Lee reckoned that, given wide enough implementation, hypertext might rescue them. He submitted a proposal in March 1989 for an "information mesh" accessible to the several thousand CERN employees. "Vague, but interesting," his boss replied. Adequately encouraged, Berners-Lee spent the next year and a half struggling to refine his idea, and also to find a suitable name. Since *mesh* sounded too much like *mess* he considered calling his system the "mine of information," but his time at CERN had taught him that a distinctive acronym could be as instrumental to a project's success as a solid concept, and MOI, meaning *me* in French, hardly embodied the notion of information sharing at the multilingual institution. He also noticed that the letter Z had gone over well in previous CERN software, so he quixotically tried narrowing his search to words starting with the almost as exotic W.

"WorldWideWeb: A Proposal," authored by Berners-Lee and his colleague Robert Cailliau in November 1990, made extensive use of *W*s while fortuitously highlighting a word much more evocative than *mine* or *mesh*. "HyperText is a way to link and access information of various kinds as a web of nodes in which the user can browse at will," the pitch asserted. "The web need not be hierarchical, and therefore it is not necessary to 'climb up a tree' all the way again before you can go down to a different but related subject. The web is also not complete, since it is hard to imagine that all the possible links would be put in by authors. Yet a small number of links is usually sufficient for getting from anywhere to anywhere else in a small number of hops." The more expansively they used the word *web*, the more compellingly the metaphor illuminated Berners-Lee's ideas. The word *mesh* would have suggested something manufactured, a product too tightly woven to stretch. The word *mine* would have had almost the opposite connotation, of something discovered, a limited resource to be stripped. The beauty of the word *web* is that it implies a spontaneous intelligence stitched through group effort. The WorldWideWeb was born as a naturally evolving ecosystem of information, the online equivalent of the food web in a glade or a pond.

Berners-Lee even believed that an information scientist might study it, as an ecologist researches a food web in the wild. "Perhaps a linked information system will allow us to see the real structure of the organisation in which we work," he suggested in his initial 1989 proposal, a plausible idea at the beginning of the web's existence. What forever changed matters was a brief posting by Berners-Lee to the alt.hypertext newsgroup on August 6, 1991, shortly after the web went global with a second node at the Stanford Linear Accelerator Center. Publicly describing his WorldWideWeb, he wrote "Collaborators welcome!" People seized on his invitation. By 2009, widely celebrated as the web's twentieth anniversary, there were more than 230 million sites, with an average of 6 million new ones being added each month. It still was a web, technically speaking—and within it one could still navigate from node to node as Berners-Lee intended—yet viewed in aggregate it had grown as nebulous as the millions of stars comprising the Clouds of Magellan.

The metaphoric use of clouds to represent vagueness is almost as old as the word. As early as 1509 Stephen Hawes was writing of "cloudes derke and termes eloquent" to describe "the arte of rethoryke." Telecom engineers in the twentieth century somewhat more whimsically sketched clouds to represent the uncharted networks into which their equipment connected. Computer vendors selling large systems such as batteries of automatic teller machines even wrote the word *cloud* in their marketing materials to designate relay centers without revealing anything about what went on inside the buildings.

The origin of cloud computing, which the media started writing about in 2006, was a sort of return to this model of business, revised and expanded to take advantage of the web. In August of that year, Amazon started a venture called the Elastic Compute Cloud, which rented use of the company's idle servers, accessed over the internet, to businesses needing additional computer power on a provisional basis. Google began offering similar "web services" at about the same time, which CEO Eric Schmidt called *cloud computing* in a widely covered speech. Other companies with computer power to spare swiftly entered the business or rebranded the services they were already offering to fit the trend (a practice aptly dubbed *cloudwashing*). Microsoft launched Azure, Juniper introduced Stratus, Sun Microsystems announced Sun Cloud, and Salesforce billed its corporate convention in San Francisco the "Woodstock of cloud computing"—hiring people to stand outside in cloud costumes—a stunt that, according to the *Wall Street Journal*, helped bump annual revenue by 44 percent.

As the cloud inflated, so did the hype. The information-technology research firm Gartner predicted that cloud computing would be "as influential as e-business." *BusinessWeek* called it "supercomputing for the rest of us." And the *Economist* approvingly quoted IBM "technology visionary" Irving Wladawsky-Berger's comparison of the cloud to the Cambrian explosion.

Missing Woodstock could be written off as a generational thing, but to be excluded from the Cambrian explosion had the ring of extinction. Companies of all kinds embraced the cloude of rethoryke. The term came to encompass not only what had once mundanely been known as Platform-as-a-Service (PaaS) but also Infrastructure-as-a-Service (IaaS) and Software-as-a-Service (SaaS), which is to say that everything from global shipping fulfillment to Gmail was a cloud service.

Responses to the cloud's swelling cloudiness have varied. The science fiction writer William Gibson, quoted in *The Industry Standard*, lauded the cloud as "numinous" and argued that its "main usefulness lies in its vagueness, like cyberspace," the infinitely adaptable neologism that he himself coined in 1984. On the other hand Larry Ellison, in his rant to financial analysts, complained, "We've redefined cloud computing to include everything that we already do." And then he compared it to the vapid chic of women's fashion.

Both of them make valid points. Cloud computing is fundamentally a case of what Ellison began describing all the way back in 1999, when he started proselytizing that information technology would increasingly work like an electric or telephone utility, running across a network to provide computation on demand (an idea that was itself an update on mainframe time-sharing in the 1960s). Still, cloud computing has picked up on all that digital technology has become since then, so that the cloud is not merely a source of petaflops, let alone a telecom relay station. The cloud is everything that the web does. In that respect the term has to be vague. If it weren't, it wouldn't be accurate.

Yet the cloud, for all of its success, has not subsumed the web linguistically or technologically. Though the metaphors seem to be in opposition, together they capture a dynamic that is the source of profound innovation. The proliferation of hypermedia—Collaborators welcome!—has produced a cyberspace in which local developments are blind to global consequences. That is the dynamic of evolution.

In Vitro Meat

Steak and chops grown from cultured muscle cells in an industrial vat.

According to the eminent seventeenth-century botanist John Parkinson, one of the plants that grew in the Garden of Eden was the vegetable lamb. Also known as the borametz, this creature resembled a young sheep in every important respect, except that it grew from a seed planted in the ground. Reports of it date back at least as far as Herodotus, and the fourteenth-century explorer John Mandeville claimed, in his notoriously unreliable *Voyages and Travel*, to have tasted one "although it were wonderful."

Only in the 1800s was the legend debunked, largely on the initiative of the British naturalist Henry Lee. (He convincingly speculated that borametz rumors began with the spread of the cotton plant, which to an untutored eye looked as woolly as a sheep.) Yet the dream of cultivating meat off the hoof, of growing muscle without the animal, was not so easily dismissed. In his 1931 book, *Fifty Years Hence*, no less a figure than Winston Churchill anticipated a time when "we shall escape the absurdity of growing a whole chicken in order to eat the breast or wing, by growing these parts separately under a suitable medium." He didn't live to see it happen. Half a century passed and vegetable meat remained as elusive as the Garden of Eden. But the technologies necessary for cultivation were evolving, quietly developing as researchers studied subjects as far afield as organ transplants and stem cells. Gradually a few laboratories, some of them funded to develop tastier astronaut cuisine for NASA, began growing potentially edible animal tissues in a bioreactor. Then everything changed again in 2008, when People for the Ethical Treatment of Animals (PETA) announced a prize of $1 million to the first person to produce "an in vitro chicken-meat product that has a taste and texture indistinguishable from real chicken flesh to non–meat-eaters and meat-eaters alike." Almost overnight in vitro meat, or at least the idea of it, was headline news: The ancient dream, newly named, went prime time.

Except that to many people, perhaps a majority, it sounded like a nightmare. The term *in vitro meat* evoked a dystopia midway between *Brave New World* and *The Matrix*. A few bloggers even suspected that PETA's contest was meant to repulse people into going vegetarian. But PETA's offer was in earnest, a pragmatic gesture that caused a "near civil war," cofounder Ingrid Newkirk confessed to the *New York Times*. Moreover the term *in vitro meat* was not PETA's invention, and the man who claims to have coined it, Willem van Eelen, is one of the technology's most steadfast advocates. Now the world's only recipient of in vitro meat patents, van Eelen started thinking about lab-grown victuals in the 1940s, after nearly starving to death in a World War II prison camp. For him *in vitro meat* was a term that evoked salvation from hunger through scientific progress.

In vitro meat thus has the strange status of being simultaneously an accolade and an epithet. It's the linguistic equivalent of a double agent. Researchers use the word to advocate their work, and Newkirk has gushed that "in vitro meat is a godsend." Critics, on the other hand, gleefully embrace the term as a phrase of readymade irony, as the *Times* food columnist Mark Bittman has done under headlines such as "Get Out Your Chemistry Sets: It's Time to Make Meat!"

Certainly there are other linguistic double agents. For instance, referring to someone as a Republican in a red state generally carries different connotations than it does in a blue state. What makes *in vitro meat* notable is that it has no inherent politics. In fact, looked at from a neutral position the term is purely descriptive.

In vitro, literally meaning "in glass," was first formally defined in the 1894 edition of Gould's Dictionary of Medicine, which noted that the term was "applied to phenomena that are observed in experiments carried out in the laboratory with microörganisms, digestive ferments, and other agents, but that may not necessarily occur within the living body." This definition is still essentially accurate—allowing for the fact that most laboratory glassware is now plastic—and it applies to in vitro meat as aptly as it does to in vitro hemolysis or in vitro ubiquitination. To say that meat is *in vitro* is simply to specify how the muscle was grown, in contrast to the *in vivo* meat sold at the local butcher shop.

Of course nobody refers to a steak or a chop as *in vivo meat*, and not only because the *in vitro* alternative is still commercially unavailable. The word *meat* is not scientific, whereas *in vitro* and *in vivo* are exclusively so. The linguistic clash calls attention to the language, freighting it with significance. And what it signifies depends on the speaker. For those who see technology as a panacea, in vitro meat evokes progress. For those who view technology as a monster, in vitro meat has the sound of frankenfood, with a built-in yuck factor—which is a serious impediment to communication, signaling a cultural divide and suggesting why such schisms are so difficult to bridge.

Can the dialogue about in vitro meat ever become a real conversation? Proponents sensitive to the opposition have suggested alternatives, such as *cultured meat* and even *victimless meat*, which is to say that they've recognized the value of marketing; opponents have come up with *meat without feet*, which is to say that they've recognized the value of rhetoric. What neither side seems to see is the virtue of discussion.

As in vitro meat becomes reality, it may have less in common with the universally acclaimed borametz than with manna, which took its name, according to Exodus, from the Israelites' question to Moses, "What is it?" and possessed, according to the Talmud, "all flavors at will." Palatable to those who desire it, repugnant to those who resist it, in vitro meat promises to be, true to its word, whatever we make of it.

QUBIT

A quantum bit.

Every man in ancient Egypt had his own standard of measurement. The standard unit was a cubit, the span from the bend of his elbow to his middle finger tip.* A natural length, as natural as the body itself, it was also a natural source of disagreement between men of different statures. To build the Pyramids, or even conduct daily business without coming to fisticuffs, the cubit had to be universalized.

And it was, more than once. The Egyptian royal cubit was one palm longer than the common cubit of 17.72 modern inches, that palm being scaled to match the hand of the reigning pharaoh. In Greece, where the cubit was adopted, it needed to be reconciled with the local fathom, based on the distance between two outstretched arms. Adjusted to measure one

---

* The etymology of *cubit* is *cubitum*, Latin for "elbow." Obviously this name wasn't used in pre-Roman times. However the lexical impulse was the same: the Egyptian hieroglyph for *cubit*, whatever the original pronunciation of the word, depicted an elbow and forearm.

quarter of a fathom (the fathom having been standardized to six Greek feet of 12.14 modern inches each), the Greek cubit spanned 18.2 inches—which is to say somewhat less than the Babylonian cubit of 19.6 inches and somewhat more than the Roman cubit of 17.49 inches—none of these measures correlating with the Hebrew, Persian, Arabic, or Mesopotamian cubits.

For the ancient trader, the range of cubit lengths must have been a source of aggravation. For the modern historian, though, these differences in definition—discrepancies in measurement preserved in written records and stone rods—are a rich source of information about ancient cultures, their bureaucratic structures, and their interactions with others. Future historians may find equivalent insight into our own time by examining our tortuous path toward standardized measurements of information.

Our most basic unit is the bit, an abbreviation of *binary digit*. The word first appeared in print in 1948, in one of the founding papers of information theory, "A Mathematical Theory of Communication," written by Claude Shannon, father of the discipline. Shannon borrowed *bit* from the statistician John Tukey, a colleague at Bell Laboratories who had a knack for coining catchy words (such as *software*), and came up with *bit* as an alternative to less attractive contractions, including *binit* and *bigit*.

Like the cubit, the bit was in use long before there was a name for it, let alone a formal definition. Even the first programmable machines, ancestors of the modern computer, were guided by a sort of binary code. These machines were looms, initially developed in the 1720s by Basile Bouchon and Jean Falcon and perfected at the beginning of the nineteenth century by Joseph Marie Jacquard. They worked by reading the arrangement of holes punched on a card with a lattice of spring-activated pins connected to hooks that would each individually lift a warp thread wherever a pin entered a hole. Cards were strung together and fed through the loom in succession, one per line of brocade. Patterns could be modified by rearranging the card deck, in much the way that a PC can be made to process words or generate spreadsheets by running different software through the same circuits. The binary code of computer software is represented by a string of ones and zeros that turn on and off switches within a microprocessor. Those switches are analogous to the spring-activated pins inside the loom: each bit was programmed into the card by punching a hole, or not punching one, in the zone touched by a pin when the program ran. On and off: binary code was, and remains, the most straightforward solution to the problem of controlling a machine.

Yet calling it a bit, and formalizing what that meant, was as profound an intellectual achievement as the stride taken in ancient Egypt from consulting body parts to creating standards of measurement. Shannon's theory allowed information to be quantified and analyzed. It became clear that determining the information content of a punch card or line of code was not just a matter of tabulating the number of bits in a sequence. It also required determining their predictability; a bit with a 75 percent probability of being on when the one preceding it is off, for instance, contains less information than one for which the chances are 50/50. Building algorithms that leverage these probabilities became the basis of data compression. Those algorithms make the communication and storage of information efficient and are the mathematical key to a worldwide digital network exponentially expanding in size from gigabytes to exabytes to yottabytes.

But what exactly is a gigabyte? Here culture bears an unmistakable imprint. Whereas the bit is a logical unit,* calling eight bits a byte is merely conventional, and counting off 1 trillion bytes in a gig is considered by computer scientists to be risible.

The word *byte* was introduced in 1956 by the IBM engineer Werner Buchholz in an internal memo about the company's first transistorized computers. As he explained in a 1962 book on the project, "Byte denotes a group of bits used to encode a character." Any number of bits could comprise a byte—an intentional misspelling of *bite* whimsically evoking the dollop of binary digits a computer could swallow at once—and for a brief time a byte was just six. But the permutations possible with six bits, 64 or 2^6, were insufficient if the character set was to include a lowercase alphabet in addition to capitals, numbers, and mathematical symbols. Accordingly some engineers started to toy with seven bits to a byte, yet the number eight appealed more to the aesthetic sensibility of mathematicians (even if nobody had any idea what to do with all 256 characters). By the early 1960s eight bits per byte was the standard.

Engineers and coders who found eight-bit bytes inconvenient invented alternative units, albeit generally adhering to the gastronomic naming scheme. Two bits were called a *tayste* or *tydbit*; four bits were a *nybble*; sixteen or eighteen bits were a *chawmp*. Thirty-two or forty-eight bits were a *gawble*, or sometimes a *dynner*, in which case a chawmp might be referred to as a *playte*. Such punning was characteristic of late-night programming humor, and in all cases except for the nybble these names were strictly internal

---

*   Logical is not the same as inevitable. In 1956 researchers at Moscow State University developed the Setun, a ternary computer with three possible values for each "trit." Trits contain more information than bits—approximately 1.58 bits per trit—and the Soviet engineers claimed that the Setun was more powerful and less expensive than a binary computer. Approximately fifty Setuns were manufactured and installed in factories and universities before production ended in 1965. An improved model, featuring six-trit "trytes," fared no better, another casualty of binary thinking.

jargon, only adding to the confusion. For clarity some computer scientists insisted on calling a byte an *octet*. The name never caught on. That the word *byte* and the eight-bit convention persisted isn't surprising, given the pragmatic culture of engineering, where legacy systems often run for decades under layers of new software. *Byte* is legacy language for a legacy standard.

Only sometimes legacy systems are brought to a breaking point. Such is the case with the prefixes added to bytes in aggregate. In the days when a thousand bytes counted as a vast amount of data, engineers noticed that the binary exponential 2^{10} was very similar to the decimal exponential 10^3, the difference between 1,024 and 1,000 being a mere 2.4 percent. So, for the sake of convenience, they took to using the familiar metric prefix *kilo-* (10^2) to refer to 2^{10} bytes. The term proved popular, and *mega-* (10^6) sounded even better, especially to computer companies trying to persuade customers of their products' power. The trouble was that the difference between 10^6 and 2^{20} is nearly 5 percent, and between 10^9 (*giga-*) and 2^{30} is more than 7 percent. In 2009 the amount of data on the internet was estimated to be 500 exabytes. At exabyte scale, the discrepancy between decimal (1,000,000,000,000,000,000) and binary (1,152,921,504,606,846,976) is on the order of 15 percent.

The International Electrotechnical Commission finally stepped in to remedy this problem in 1999, with a brand new set of names for binary multiples. Thereafter 2^{10} bytes were officially a kibibyte (*kibi-* being a contraction of *kilobinary*). Larger multiples followed the same scheme: *mebibyte, gibibyte, tebibyte, pebibyte, exbibyte*. By 2005, with the internet surging past the exabyte (and exbibyte) mark, even bigger prefixes were called for, and the IEC responded with *zebi-* and *yobi-*, the binary equivalents of *zetta-* and *yotta-*. A yobibyte is 1,208,925,819,614,629,174,706,176 bytes.

Five years later a yobibyte has yet to be manifest in all the world's circuitry. Not yet useful, the word is a symbol of technological optimism, the belief that there will always be exponentially more of what was once unfathomably much. This is the promise of Moore's Law, the long-standing prediction that the number of transistors in a microprocessor will double every two years.

Honest technologists doubt that Moore's Law will hold true far into the future. Electronic components can be shrunk only so much before they reach quantum scale and are subject to quantum interference—unless of course the computer becomes a quantum machine in its own right. In that case bits will no longer each be limited to a simple binary value of zero or one. A quantum bit can be put in a quantum superposition in which it's both at the same time. Moreover multiple bits can be entangled—another uniquely quantum phenomenon—such that each new bit exponentially increases the power of the computer. One of the physicists developing the underlying information theory for this, Benjamin Schumacher, punningly named these quantum bits in a 1995 paper. He called them *qubits*.[*]

Obviously qubits have little in common with cubits, Roman or Greek or Mesopotamian. However, like Claude Shannon's bit, they do share something with the standardized Egyptian forearm. Even the most optimistic engineers don't anticipate commercial quantum computers in the next ten years, and laboratories still struggle to control even a one-qubyte computer, yet the mere act of designating a quantum unit of information is providing a powerful new way to take the measure of the quantum realm.

---

[*] Schumacher later described the circumstances of his naming to the *Dallas Morning News* correspondent Tom Sigfried, who recounted what Schumacher said in his 2000 book, *The Bit and the Pendulum*. The term originated in a 1992 conversation with the physicist William Wootters: "We joked that maybe what was needed was a quantum measure of information, and we would measure things in qubits was the joke, and we laughed. That was very funny. But the more I thought about it, the more I thought it was a good idea. . . . I thought about it over the summer and worked some things out. It turned out to be a really good idea."

Gene Foundry

A laboratory that produces custom DNA for purposes of bioengineering.

As the nineteenth century was the age of iron and the twentieth belonged to silicon, the present century will be identified with carbon. CO_2 is the iconic greenhouse gas, imprinted on our vocabulary with talk of carbon footprints and allowances and offsets. For synthetic biologists, however, the carbon debacle has counterintuitively positioned this debased element as our savior. The future they foresee will supplant grimy factories of concrete and steel with clean colonies of living cells. To use the terminology of Freeman Dyson, gray technology will be replaced by green.

Among the most celebrated physicists of the twentieth century, Dyson has become one of the foremost promoters of synthetic biology, a field that technologically is to genetic engineering what genetic engineering is to crop cultivation. Conceptually the distinction is even more radical than that: whereas genetic engineering merely modifies preexisting creatures more precisely than selective breeding, synthetic biology aims to fabricate entirely new organisms from nonliving materials. Unconstrained by genetic history, these artificial life forms can be intelligently designed to produce fuels or pharmaceuticals with unprecedented efficiency. The sheer audacity of synthetic biology lends itself to hyperbole, aptly captured in a 2007 *Nature* editorial: "For the first time, God has competition."

Yet the language of synthetic biology, also known as bioengineering, hardly bespeaks a cosmic paradigm shift. DNA constructed at a so-called gene foundry gives specialized function to a generic cell referred to as a *chassis*. A Victorian industrialist would have no trouble following the metaphoric language. He might even find work as a bioengineering consultant: the quaint iron age phrasing reflects the old-fashioned framework underlying this brave new discipline.

After all, radical as artificial life may be philosophically—and significant as it may be environmentally—it's technically just a strenuous construction project, with manufacturing challenges akin to building a bridge or a steam engine. That may explain why one of the most successful synthetic biologists working today, the Stanford University professor Drew Endy, trained as a civil engineer. Together with colleagues at the Massachusetts Institute of Technology, Endy has methodically approached synthetic biology as a problem of developing reliable building blocks and assembly protocols.

These efforts have been consolidated in a Registry of Standard Biological Parts, which lists thousands of "biobricks," defined by the Registry as components "that can be mixed and matched to build synthetic biology devices and systems." Bioengineers seeking to construct a cell with novel utility can browse the biobrick collection "by type, by function, by chassis and by standard," much as a mechanical engineer might browse a catalogue of standard pistons and gears. For instance, to make a microbe convert isoamyl alcohol to the odor isoamyl acetate—artificial banana scent—the Registry provides a 1,581-letter DNA sequence:

```
atgaatgaaatcgatgagaaaaatcaggcccccgtgcaacaagaatgcctgaaagagatgattca
gaatgggcatgctcggcgtatgggatctgttgaagatctgtatgttgctctcaacagacaaaact
tatatcgaaacttctgcacatatggagaattgagtgattactgtactagggatcagctcacattagcttt
gagggaaatctgcctgaaaaatccaactcttttacatattgttctaccaacaagatggccaaatcat
gaaaattattatcgcagttccgaatactattcacggccacatccagtgcatgattatatttcagtatta
caagaattgaaactgagtggtgtggttctcaatgaacaacctgagtacagtgcagtaatgaagcaaat
attagaagagttcaaaaatagtaagggttcctatactgcaaaaatttttaaacttactaccactttgact
attccttactttggaccaacaggaccgagttggcggctaatttgtcttccagaagagcacaca
gaaaagtggaaaaaatttatctttgtatctaatcattgcatgtctgatggtcggtcttcgatcca
cttttttcatgatttaagagacgaattaaataatattaaaactccaccaaaaaaattagatta
cattttcaagtacgaggaggattaccaattattgaggaaacttccagaaccgatcgaaaaggtgata
gactttagaccaccgtacttgtttattccgaagtcacttctttcgggtttcatctacaatcattt
gagattttcttcaaaaggtgtctgtatgagaatggatgatgtggaaaaaaccgatgatgttgtcaccga
gatcatcaatatttcaccaacagaatttcaagcgattaaagcaaatattaaatcaaatatccaagg
taagtgtactatcactccgttttttacatgtttgttggtttgtatctcttcataaatgggg
taaattttttcaaaccattgaacttcgaatggcttacggatattttttatcccccgcagattgccgctca
caactaccagatgatgatgaaatgagacagatgtacagatatggcgctaacgttggatttattgact
tcacccctggataagcgaatctgacatgaatgataacaaagaaaatttttggccacttattgagcac
taccatgaagtaatttcggaagctttaagaaataaaaagcatctccatggcttagggttcaatata
caaggcttcgttcaaaaatatgtgaacattgacaaggtaatgtgcgatcgtgccatcgggaaaa
gacgcggaggtacattgttaagcaatgtaggtctgtttaatcagttagaggagcccgatgccaaatat
tctatatgcgatttggcatttggccaatttcaaggatcctggcaccaagcattttccttgggtgtttgt
tcgactaatgtaaaggggggatgaatattgttgttgcttcaacaaagaatgttgttggtagtcaa
gaatctctcgaagagctttgctccatttacaaagctctcctttttaggcccttaataa
```

A bioengineer need simply email this formula to a foundry—where the bases adenine, cytosine, guanine, and thymine are chemically joined to produce custom genes—and then load the made-to-order DNA onto a cellular chassis, such as the bacterium e. coli, to have a microscopic banana-odor factory.

The uses of banana scent are admittedly limited, yet producing isoamyl acetate is technically akin to producing ethanol fuel, or even the malaria drug artemisinin (as the UC Berkeley bioengineer Jay Keasling has done). As Endy explained in a 2005 *Guardian* article, the overarching goal is to "produce components that are insulated from one another, that are designed to be easy to be put together and then to behave in ways that you expect."

Insulation is the key idea. And just as the design of complex systems is insulated from DNA minutiae by the assembly of self-contained biobricks, the grand ambition of bioengineering is insulated from biological details by the use of all-encompassing metaphors. Speaking of a *chassis*, bioengineers can bracket the question of whether that chassis ought to be an e. coli cell or mycoplasma genitalium. More broadly the industrial metaphor is a constant reminder that the impetus of bioengineering is engineering, not biology. Practitioners are prompted to pursue the distant goal of truly artificial life, in which every aspect of the cell can be selected because it's needed for the job and performance is optimized through complete knowledge and total control of the whole system. The metaphoric language may even help guide synthetic biologists to this end by inspiring broad organizational decisions such as the institution of the Registry of Standard Biological Parts, which emphasizes assembly-line consistency over bespoke craftsmanship. The question is whether the industrial metaphor will also burden synthetic biology with unanticipated limitations.

It's happened before, albeit in a rather different arena. Four decades after researchers at Xerox came up with the so-called desktop metaphor as a user-friendly interface for the personal computer, operating systems are still designed in terms of metaphoric documents and folders. The original justification for the desktop metaphor was unassailable: computers were alien to most people and would find widespread acceptance only if interaction with the machine could be made intuitive. For that to happen, the digital domain had to feel familiar, and since PCs were intended for the workplace, that meant loading up the screen with virtual office supplies. And it worked, if not for Xerox, then for Apple, which found widespread success with the Macintosh in 1984, and for Microsoft, which appropriated Apple's desktop aesthetic for the Windows operating system a year later. Ever since then people have complained that the metaphor is constraining.

One of the earliest criticisms was published in *MacWeek* on March 21, 1989. "In the eyes of some developers and users, the desktop metaphor has worn thin," the magazine reported. "As anyone who's been around the desktop knows, the consistent metaphor is in fact inconsistent. Metaphors get mixed, with windows residing on top of the desktop, and messages get garbled, with the trash can being used to delete a file and to eject a disk." The grievance was pretty much the same in 1997, when *InfoWorld* editorialized, "It's time to put the desktop metaphor in the circular file and dream a little," and in 2001, when *Technology Review* called the desktop "an unmanageable mess" and quoted the computer scientist David Gelernter saying, "The desktop is dead." He was wrong, of course, and six years later his eulogy was still premature, his prediction still unfulfilled, when MIT Press published an anthology titled *Beyond the Desktop Metaphor.* The book dwelled mostly on the past, during which multiple alternatives had been introduced and abandoned, such as Lifestreams (which organized files chronologically in a metaphoric diary) and Microsoft Task Gallery (which organized files spatially in a three-dimensional labyrinth of metaphoric rooms). A convincing explanation for the repeated failure was given by John Bowman on *CBC News* in 2006: "Even critics of the desktop metaphor concede that it's become so ubiquitous, so natural, that adopting a new one would be like learning Esperanto—a good idea in theory, but for most people not worth the trouble." Naturally Bowman was reporting on another would-be replacement, BumpSpace, which took metaphoric inspiration from computer games.

Might the industrial metaphor cripple innovation in synthetic biology as the desktop metaphor has done in personal computing? Already the language is showing equivalent strain, with biobricks being fabricated at foundries and loaded onto chassis.* This is not so much a sign that bioengineers are insane, as an indication that the metaphor has become entrenched, sufficiently established to hold its own against common sense. And such language leads bioengineers to take for granted that synthetic biology is to be achieved by corporations and universities at an industrial scale. Factories may be comprised of living cells, but those cells will still reside in factory buildings.

Synthetic biology is new enough that we can still imagine alternate paths. Freeman Dyson has presented one of the most visionary in an essay published in the *New York Review of Books*. While acknowledging big industrial applications, such as earthworms engineered to extract aluminum from clay, he predicts "that the domestication of biotechnology will dominate our lives during the next fifty years at least as much as the domestication of computers has dominated our lives during the previous fifty years." By *domestication* Dyson means that synthetic biology will be an activity as casually undertaken as orchid cultivation or dalmatian breeding, with genes freely traded between species through the cultural nexus of humans. Thus inefficient Darwinian evolution will be replaced by a sort of interspecies open-source mashup.

---

* Needless to say, cars are not built with masonry, and masonry is not manufactured at an ironworks.

Whether this biopunk future is preferable to industrial bio-engineering is debatable. Either may be our salvation, and both have the potential to go awry, causing more ecological damage than any amount of carbon dioxide. Language can't talk us out of environmental Armageddon. But at least we can have some say in our fate—on our desktops and on our planet—by mastering our metaphors rather than being mastered by them.

Memristor

A resistor with memory.

The capacitor was discovered in 1745 by Ewald Georg von Kleist, whose encounter with a generator and a jar of water shocked him so severely that he declared himself unwilling to repeat the experience "for the kingdom of France." The resistor announced itself to mankind somewhat less dramatically in 1827, followed by the inductor in 1831. For the next 140 years these three components were considered the basic elements of electronics. Each accomplished what the others could not, even in combination, and together they gave engineers rudimentary control over electromagnetism. The capacitor linked charge and current, the resistor, current and voltage, and the inductor, current and flux. Later innovations, most notably the invention of transistors in 1947, would vastly expand the capability of electronics and even more incredibly stretch our expectations, yet everyone remained satisfied with the three old "passive" elements. If any more existed there simply was no need to find them.

Then along came a young engineer named Leon Chua, who, unusual for someone in his profession, had an Aristotelian turn of mind. Instead of asking himself what could be *done* with capacitors and resistors and inductors, he sought to define what they *were*. His definitions, expressed in abstract terms of charge and current and voltage and flux, suggested to him an incomplete pattern, like a crossword puzzle with all but one word filled in. In 1971 he predicted the existence of a missing link between flux and charge. He gave it a name. He called his component a *memristor*.

Still, it was only a placeholder, since nobody had ever seen one or cared about manufacturing them. His mathematical reasoning was elegant, acknowledged those who bothered to follow it, but engineers were much more excited by his 1983 invention of a simple circuit that behaved chaotically (in the formal mathematical sense), with obvious applications in computing and security. The circuit was named in his honor, making him a very minor celebrity.

By 2001 even Chua had more or less forgotten about his hypothetical memristor, but a Hewlett-Packard researcher named Greg Snider was reminded of the oddly named component when the molecular electronics he and his colleagues were developing began behaving unaccountably. The puzzling behavior had also been encountered by others, and was all the more troubling because beneath it lurked the fear of engineers everywhere: the bottoming-out of Moore's Law as materials assumed different properties at atomic scale. Snider showed Chua's thirty-year-old paper to his collaborator Stan Williams, who promptly changed the focus of their research.

The memristor gets its name from an unusual characteristic. Resistance increases as current flows through in one direction and decreases when the current is reversed. If current is cut off, the memristor preserves, or "remembers," the level of resistance reached when current last passed through it. The reason for this is that voltage exerts a slight force, subtly redistributing the atoms in the crystal structure of a semiconductor, altering its degree of resistance. In a large crystal (such as those used in electronics circa 1971) the effect is insignificant. As semiconductors are scaled down, though, the effect becomes noticeable, one of the myriad problems faced by chip designers trying to shrink transistors. Except Williams didn't see it as a problem. To him it looked like a solution.

As Chua had anticipated in his paper, the memristor's memory could be used to store data, retrieved by measuring the component's resistance. Moreover the memory was "nonvolatile," meaning that it was preserved without power. Given the rate of processors and the size of chips in the 1970s, all of this was of academic interest. But over the thirty years Chua's paper moldered and the additional seven years that Williams and colleagues spent building a reliable memristor from doped titanium dioxide before presenting their findings in *Nature*, the need for ultracompact nonvolatile memory increased exponentially. The first practical circuits are now in development, and a more radical application is already under consideration. Unlike transistors, which are either on or off, memristors aren't inherently digital, since resistance can be adjusted gradually and set to any level. They are naturally suited to analog computing, equivalent to the function of the brain, the processing power and efficiency of which remain impressive even to computer scientists. Many see analog computing as an alternative to transistor meltdown, an escape from the miniaturization endgame. By this interpretation, the ultimate redemption for Moore's Law is the humble memristor.

It is almost too good a story, a narrative arc worthy of Aristotle's *Poetics* (as popularized by the Hollywood screenwriter Syd Field): the memristor, a missing link awaiting discovery since 1831 and hypothesized for no particular purpose in 1971, becomes useful at the moment it becomes viable, and becomes viable for the same reason prior electronics are no longer feasible. That's the plotline echoed by sources ranging from the *New York Times* to *Scientific American* to *Gizmodo*. Is it reliable? Or are we seduced by the phrasing?

One hint that we're besotted comes from Chua, who has returned to his old work and begun researching *memcapacitors* and *meminductors*. Both undoubtedly have the potential to become useful components in future computers. But wasn't there supposed to be only *one* missing link in the interplay between charge and current and voltage and flux? Rather than the fourth element, the memristor appears to be the first member of a new category of passive components with memory. Many of Aristotle's definitions, eminently reasonable, also faltered upon further observation. And most of his explanations proved too perfect for the real world.

However, Chua would not be pioneering memcapacitance and meminductance had he not also been seduced by the compulsion to seek aesthetic order in electronics. The role that words and stories play in discovery may be deceptive, but the deception is essential unless we believe what we say more than what we see.

The story of the memristor, too good to be true, is the story of technological discovery: an unfounded imaginative leap, redeemed by serendipity, driven to market on the basis of its own mythology.

PART III

COMMENTARY

On the night of November 4, 1811, several disgruntled weavers smashed some mechanical looms in a small English village. They claimed to be led by a Captain Ludd. Their story swiftly spread across the countryside, spurring other craftsmen to take similar action against the new machines and mills threatening their livelihood. Over the next fourteen months workers throughout the English Midlands rebelled against the Industrial Revolution. Some eleven hundred knitting machines were destroyed in Nottingham. Twelve factories were raided in Lancashire, and in Yorkshire a manufacturer was shot dead. Sometimes a general, or even a king, Ludd was always in the lead.

It helped that he didn't exist. While his followers were captured and brought to trial, many of them hanged under a new law against Luddism, Ned Ludd could not be killed. And he still survives, at least in the English language, as the prototype for the term *luddite*.

The meaning of *luddite* depends on who's using the word. Call yourself a luddite and you're probably proclaiming your opposition to technological development for its own sake, with the implication that technology is dehumanizing. Call someone else a luddite and it's more likely that you're declaring him or her to be incompetent, with the implication that humans are technological beings. Such is the case with many words that comment on society, such as *crowdsourcing*, which may be complimentary or insulting, depending on the context. In fact that resilience is part of what makes such words endure, while narrower terms such as *bacn* are swiftly forgotten.

Commentary can thrive in a word, as long as it is as elusive as Captain Ludd.

Bacn

Spam by personal request.

Seldom has the arc of a neologism been so visible. On the afternoon of August 18, 2007, standing at the PodCamp Pittsburgh registration desk, Tommy Vallier, Andy Quale, Ann Turiano, Jesse Hambley, and Val and Jason Head—all participants in the city's annual social media conference—were having a conversation about Canadian bacon. Vallier informed the group that peameal bacon was an alternate name for the breakfast meat, leading others to comment that *peameal* sounded like *email*. This coincidence in turn reminded them of a prior discussion about all the automatic email notifications they received daily, from Google news alerts to Facebook updates, which were becoming almost as distracting as spam. They decided it was a problem, and their banter about peameal and pork suggested a name. Since the notifications were a cut above spam—after all, these updates had been requested—they dubbed this "middle class" of email *bacn*.

The following day the six PodCampers held a spontaneous group session with several dozen of their fellow social media mavens, who were swiftly won over by the jokey name and ironic spelling (a play on sites such as Flickr and Socializr then popular). The web address bacn2.com was acquired—bacn. com was already taken by a bacon distributor and bacn.org belonged to the Bay Area Consciousness Network—and a droll public service announcement explaining the time-wasting dangers of bacn was promptly posted on YouTube. What happened next was best explained by PodCamp's cofounder Chris Brogan to the *Chicago Tribune* five days later. "The PodCamp event was about creating personal media," he said, "so 200-something reporters, so to speak, launched that story as soon as they heard it."

The term was written up on hundreds of personal blogs, bringing it into Technorati's top fifteen search terms and leading Erik Schark to muse on *BoingBoing* that the spread of bacn showed "the ridiculous power of the internet." Schark also listed the mainstream media that had covered it, including *CNET*, *Wired*, and the *Washington Post*, where Rob Pegoraro complained about the name: "Bacon is good," he opined. "Why wouldn't you want bacon?"

That peeve was echoed by many others on the web, some of whom suggested alternative meats, such as bologna. A heftier criticism was leveled on National Public Radio by Bruno Giussani, the European director of the TED Conferences. "I don't buy it," he said. "So five or six geeks meet at a conference, start tossing names around, and then pretend to have identified a new trend."

Giussani's criticism aired on August 29, eleven days after PodCamp registration, the only skeptical voice in a piece touting bacn as "the e-mail dish du jour." But in truth the term already had a leftover quality, with the hoi polloi punning about *quakn* (bacn from geologists) and *rakn* (bacn from gardeners) on blogs read by two or three friends. With the last addition to the bacn website dating back to August 23—advertising a T-shirt reading "Bacn: Email you want, but not right now" priced at $23.90—even the originators appeared to have lost interest in their cause.

Bacn is not the only dead-end terminology to take *spam* as a point of departure. In 2004, for instance, *spit* made a brief appearance in the media as an acronym for "spam over internet telephony." *Spit* was coined by Qovia, an internet startup that developed software to block the new menace, though there weren't yet any known cases of it. "Spit isn't much of a problem now," reported *USA Today*. "'But it will be,' says [Qovia marketing VP] Pierce Reid." Three years later, with spit still hypothetical, Qovia was bought and dismantled by Cisco.

Even *spim* (instant message spam), which has the advantage of actually existing, has hardly become commonplace. Or rather, the sheer volume of spim has led most people to see it as part of the broader problem and simply to categorize IM ads for cheap Vi@gra and free software as spam.

Spam, spam, spam, spam, spam, spam, spam. The word is one of the most widely cited on the web (more even than *virus* and *worm*, according to Google). Moreover, the term has come to prominence without the social networking behind *bacn*, the corporate marketing backing *spit*, or even the predefined meaning underlying *spim*. On the contrary, *spam* evolved gradually, shifting in connotation with changing technologies, and persisting despite ongoing legal opposition from a Fortune 500 company.

One urban legend claims that *spam* began as a contraction of the Esperanto phrase *senpete alsendita mesago*, which can be translated into English as "a message sent to someone without request," but in fact the origin is considerably less highbrow.

The word alludes to a satirical 1970 *Monty Python's Flying Circus* skit on the unsavory subject of Spam luncheon meat. The three-and-a-half-minute vignette involves a woman trying to order breakfast in the Green Midget Café, which serves almost every dish with Spam. There's egg and Spam, egg bacon and Spam, Spam bacon sausage and Spam, Spam egg Spam Spam bacon and Spam, Spam Spam Spam Spam Spam Spam baked beans Spam Spam Spam, and lobster Thermidor à crevette with a Mornay sauce served in a Provençal manner with shallots and aubergines garnished with truffle pâté, brandy, and with a fried egg on top and Spam. "Have you got anything without Spam?" the customer finally asks, to the consternation of the waitress, while at a neighboring table some other patrons (who happen to be helmeted Vikings) begin to sing about the virtues of Spam. The remainder of the skit, during which the word *Spam* is uttered 132 times, mixes the woman's effort to order a dish without Spam with the waitress's struggle to keep the Spamophilic Vikings under control. Of course the Vikings prevail, ending the act by overpowering all other conversation with their ode to "Spam spa-a-a-a-am Spam Spa-a-a-a-am Spam. Lovely Spam! Lovely Spam!"*

The skit turns on the ubiquity of Spam in England combined with its unpopularity, both legacies of World War II, when Hormel shipped some 150 million pounds of the canned luncheon meat overseas to meet rationing needs, and soldiers facing three Spammy meals a day dubbed it "meatloaf without basic training." If the war made Spam iconic, Monty Python

---

* The Viking song bears more than a passing resemblance to a 1940 Spam jingle, one of the first used in advertising: "Spam Spam Spam Spam. Hormel's new miracle meat in a can. Tastes fine, saves time. If you want something grand, ask for Spam!" The jingle was part of a massive marketing effort that began in 1937, when Hormel rebranded its unpopular Spiced Ham with a name proposed by the brother of a Hormel vice president. He was paid $100 for his suggestion. The upswing on product sales was almost immediate.

gave it a cult following. This was especially the case with computer geeks, who prided themselves on their ability to recite whole Monty Python routines from memory and to converse almost entirely in arcane *Flying Circus* references. In the late 1980s they started using *spam* to signify unwanted communication.

The first appearance of the word was in MUDs, the "multi-user dungeons" that were created to support online role-playing games but often also served as chatrooms. There the conversational efforts of unwanted visitors were blocked out by typing *spam* repeatedly until no other discussion was possible, much as happened when the Vikings began singing in the Green Midget Café. From there it was a small step for gamers to start using the word *spam* as a verb to describe this sort of behavior. As the veteran MUDer Patrick J. Wetmore explained on the rec.games newsgroup in 1990, "The verb 'to spam' would be to send lots and lots of useless information (in particular, the word 'spam') over and over to someone, thus scrolling their screen with lots and lots of lines of 'spam spam spam spam spam spam' etc. It has been generalized to mean sending lots of crap to servers as well as people you want to annoy the hell out of." Wetmore's explanation, which traced the term back to Monty Python, was written in answer to a fellow MUDer's question about the meaning of *spamming*, but his real motivation was to challenge a response posted the previous day erroneously claiming that the etymology was the same as for *trashing*, i.e., "Putting so much load on the system that it crumbles into a pile of trash / Spam / other useless substance." To Wetmore this was heretical. "A man who doesn't know his Monty Python is not a man worth talking to," he wrote, summing up the MUDer ethos in a sentence.

Of course none of this was spamming in the sense that we now know it. That sort of activity can be independently traced back to 1971, when an MIT systems administrator sent an antiwar message to everyone with an address on the Compatible Time Sharing System asserting, "THERE IS NO WAY TO PEACE. PEACE IS THE WAY." Seven years passed before the next recorded case, this time decidedly more commercial. A Digital Equipment Corporation computer salesman named Gary Thuerk sent a mass email to six hundred people on ARPANET, the government-run predecessor to the internet, informing them, "DIGITAL WILL BE GIVING A PRODUCT PRESENTATION OF THE NEWEST MEMBERS OF THE DECSYSTEM-20 FAMILY." It was probably the most successful spam of all time. Whereas an estimated million emails must now be sent to garner fifteen trifling purchases, his six hundred messages resulted in at least twenty sales at $1 million apiece. The only trouble was that almost everyone who didn't buy a mainframe from Thuerk condemned him for "flagrant violation" of ARPANET. Response was so harsh that no one dared use the network as a sales tool again.

But the internet was another matter. In 1993 the law firm Canter & Siegel posted a notice on an estimated six thousand Usenet newsgroups advertising their dodgy immigration services: "Green Card Lottery—Final One?" A lot of people understandably considered that useless information—and found it annoying as hell that the law firm posted it again and again. They began calling it spam. At last there was a phenomenon worthy of the word. Both took off as one.

Spam is now considered the greatest strain on the internet and the most costly problem. Statistics vary, but approximately 94 percent of the 247 billion emails sent daily in 2009 were spam. Compared to these numbers, the 7 billion cans of Spam luncheon meat sold since 1937 seem barely worthy of mention.

As spam has threatened to eclipse Spam, Hormel has fought hard to protect its trademark. While the company has been opportunistically friendly to Monty Python's satire—even producing a commemorative Golden Honey Grail Spam to coincide with the 2005 musical *Spamalot*—corporate attorneys have systematically brought trademark infringement suits against manufacturers of spam filters such as SpamBop and SpamArrest. The corporate reasoning has been posted on the Spam website: "Ultimately, we are trying to avoid the day when the consuming public asks, 'Why would Hormel Foods name its product after junk email?'" Alas that day may already have passed. Courts have consistently rebuffed Hormel because, to quote a ruling by the European Office of Trade Marks and Designs, "the most evident meaning of the term SPAM for the consumers . . . will certainly be unsolicited, usually commercial email, rather than a designation for canned spicy ham."

Still, bacon manufacturers need not worry about something similar happening to them. In fact the roundabout way in which *spam* entered the lexicon helps to explain why neologisms seldom endure, even if initially popular. Whereas *bacn* was aggressively publicized, *spam* was quietly nurtured in a small community. As a sort of code, signifying the status of those who used it, it was protected, all the more so given the group's reverence for Monty Python. The transition into the mainstream was inadvertent, a consequence of Usenet's broadening audience and the resulting exploitation of it by advertising outsiders. By the mid-1990s few people using the word *spam* had any idea that it referred to the Monty Python skit, but almost all knew of the reviled luncheon meat. The emotional resonance was right, and the obscure history gave the word an authenticity, and an authority, that impressed it on people's vocabularies.

Bacn bears closer linguistic resemblance to *brrreeeport*, a word introduced by the widely read tech blogger Robert Scoble on February 14, 2006, with the following explanation: "Here, let's play a game. Everyone in the world say 'brrreeeport' on your blog." Many people did, enough to momentarily make a nonsense word one of the world's most blogged-about topics, topping even the chatter about Vice President Dick Cheney, who'd shot a friend on a hunting trip the previous day. *Brrreeeport* was really a social psychology experiment in the tradition of Stanley Milgram, a somewhat disingenuous demonstration of how swiftly memes travel on the internet and how willing bloggers are to become conduits. More specifically *brrreeeport* was a word not to be used, but to be observed. *Bacn* follows this pattern. "All we wanted to do was to draw awareness to the situation," Tommy Vallier told NPR on August 29, 2007, as the term began to fade.

There is a precedent for words with a mission. For instance, the term *scofflaw* was coined in 1924 as the winning entry in a contest held by the Prohibitionist Delcevare King, who offered $200 for a word that would "stab awake the conscience" of drinkers who flouted the 18th Amendment. The internet is an ideal medium for this proactive variant on language, made to strike fast and furiously, vanishing as rapidly as it appears. There will be more. *Guerrilla words*, we might call them, were we not savvy to the fate of neologisms.

Copyleft

Protection against copyright protection.

Developing an open-source alternative to the UNIX operating system in the early 1980s, the master hacker Richard Stallman faced a dilemma: if he put his new GNU software in the public domain, people could copyright their improved versions, undermining the open-source cycle by taking away the freedoms he'd granted. So Stallman copyrighted GNU himself, and distributed it, at no cost, under a license that arguably was to have greater impact on the future of computing than even the software he was striving to protect. The GNU Emacs General Public License was the founding document of the copyleft.

The word *copyleft* predated Stallman's innovation by at least a couple of decades. It had been used jestingly, together with the phrase "All Rights Reversed," in lieu of the standard copyright notice on the *Principia Discordia*, an absurdist countercultural religious doctrine published in the 1960s. And in the 1970s the People's Computer Company provocatively designated Tiny BASIC, an early experiment in open-source software,

"Copyleft—All Wrongs Reserved." Either of these may have indirectly inspired Stallman's phrasing. (He first encountered the word *copyleft* as a humorous slogan stamped on a letter from his fellow hacker Don Hopkins.) Stallman's genius was to realize this vague countercultural ideal in a way that was legally enforceable.

That Stallman was the one to do so, and the Discordians weren't, makes sense when one considers his method. His license stipulated that GNU software was free to distribute, and that any aspect of it could be freely modified *except* the license, which would mandatorily carry over to any future version, ad infinitum, ensuring that GNU software would always be free to download and improve. "The license agreements of most software companies keep you at the mercy of those companies," Stallman wrote in the didactic preamble to his contract. "By contrast, our general public license is intended to give everyone the right to share GNU Emacs. To make sure that you get the rights we want you to have, we need to make restrictions that forbid anyone to deny you these rights or to ask you to surrender the rights." Freedom was paradoxically made compulsory. With GNU Emacs Stallman effectively hacked the legal code, making copyright law accomplish the opposite of what had been intended. And he did it with a programming trick. Specifically his General Public License adapted a technique that computer scientists refer to as *recursion*.

One simple example of recursion can be found in the GNU name, which is an acronym for *Gnu's not Unix*. However, recursive loops can generate more than philosophical gags when written into computer code. Looping is the link between simple counting and complex mathematical functions: the self-referential nature of recursion is what makes a calculator into a computer. Stallman applied this powerful circularity to intellectual property law, such that a copyright protected GNU from the protections of copyright. According to the U.S. Constitution, the purpose of copyright is "to promote the Progress of Science and useful Arts, by securing for limited Times to Authors and Inventors the exclusive Right to their respective Writings and Discoveries." Believing that exclusive rights thwarted the Progress of Science and the useful Arts—that creativity was fostered by cooperation—Stallman and his GNU collaborators made the Constitution amend itself.

For obvious reasons more traditional beneficiaries of copyright protection were unamused by Stallman's constitutional hack. Most notably Microsoft denounced the scheme as "viral," and there was a sense in which the license resembled a computer virus, embedding itself in whatever came into contact with GNU. (One clause mandated, "The whole of any work that you distribute or publish, that in whole or in part contains or is a derivative of GNU Emacs or any part thereof . . . be licensed at no charge to all third parties on terms identical to those contained in this License Agreement.") But what truly proved viral was the copyleft concept, for which the word *copyleft* served as shorthand, often accompanied by the graphically powerful copyleft emblem, a simple reversal of the encircled copyright symbol.

Over the decades *copyleft* has become a catchphrase for everything from CrimethInc's radical N©! license, granting all rights to individuals but none to corporations, to the mainstream suite of licenses developed by Creative Commons, including ShareAlike and NoDerivs, a range of options to suit every liberal temperament. Some licenses, such as the one used by Wikipedia, have evolved directly from the original GNU template. Other licenses, such as Hactivismo's "enhanced-source" HESSLA contract, seek to broaden the political reach of copyleft beyond intellectual property, as HESSLA does by including a clause forbidding users "to violate or infringe any human rights or to deprive any person of human rights, including, without limitation, rights of privacy, security, collective action, expression, political freedom, due process of law, and individual conscience." Still other licenses, such as Pirate Bay's Kopimi, just offer provocative terms for putting material into the public domain.

Within this thicket there have been epic struggles, many of them generational. (Young executives who think nothing of contributing free content to Wikipedia in their spare time are scared by Stallman's "free software" rants and long scraggly hair, leading open-source pragmatists to ban him from conferences.) Yet all of the contradictory intentions and conflicting rhetorics have proven insignificant compared to the aggregate effect of the copyleft. Even as the scope of copyright law has been extended by Congress, the notion that intellectual property rights are absolute has met increasing opposition, and the opposition has grown increasingly organized. The reason is that the copyleft rubric—building on the subversively affirmative turn of the first GNU license—has given a positive identity to what otherwise would merely have been isolated shouts of frustration.

This positive identity has been bolstered by positive results, such as Linux, which was built on GNU, and demonstrates that open-source collaboration can produce software as effectively as proprietary innovation. Developed under the umbrella of Stallman's General Public License, Linux is proof-of-concept that the "Progress of Science and useful Arts" can at least in some cases be promoted without "securing for limited Times to Authors and Inventors the exclusive Right to their respective Writings and Discoveries," and that the products of collaboration may sometimes arguably be superior. The language of the GNU license simulated a legal system in which intellectual property laws were opposite those of society at large, and the output of that simulation, from Linux to Wikipedia, has given copyleft arguments real-world traction.

William Safire noted that *copyleft* is lexically akin to *software*, since both words were coined by flipping the meaning of existing language. The term *software* was first used by the statistician John Tukey in 1958 to emphasize that the latest electronic calculators were not mere "'hardware' of tubes, transistors, wires, tapes and the like," but also depended on "interpretive routines, compilers, and other aspects of automative programming." Tukey's word captured an emerging distinction, and the dichotomy helped to launch the industry that was to become the foundation for the copyleft debate. In a real sense software spawned the copyleft. However, the differences between the words *software* and *copyleft* are stronger than their similarities. Whereas *hardware* and *software* are mutually reinforcing, *copyright* and *copyleft* are antagonistic. And as *copyleft* is only meaningful by way of contrast, the copyleft will have succeeded completely—paradox of paradoxes—only when the word is obsolete.

GREAT FIREWALL

The Chinese internet.

The Chinese government declared 1996 the Year of the Internet. There wasn't much to it: only one person in ten thousand was connected—at a modem speed of 14.4 kilobits per second—and 86 percent of the population had never encountered a computer. Even in universities email was still a novelty, haltingly introduced in 1994. Yet in one respect China was the most advanced nation on the planet. Using equipment supplied by Sun Microsystems and Cisco, the Chinese Public Security Bureau had corralled the entire country, all 3,705,000 square miles, within a *fanghuo qiang*, or firewall.*

The firewall promised to make the internet safe for autocracy. All online communication could be monitored, at least in

---

* A firewall isolates a local network from the internet. Named after the stone or concrete walls isolating sections of a building to prevent the spread of fires, firewalls are typically used by companies as a security measure against infiltration by hackers and also sometimes as a means of monitoring employee interaction with the external world. The *fanghuo qiang* was constructed using the same basic technology.

principle, and access to any website could be denied. On February 1, 1996, Premier Li Peng signed State Council Order 195, officially placing the government "in charge of overall planning, national standardization, graded control, and the development of all areas related to the internet," and expressly forbidding users "to endanger national security or betray state secrets." Enforcement was arbitrary. Discipline was imposed by the dread of uncertainty. This was an inevitability, since the Public Security Bureau couldn't possibly watch all online activity within China, let alone block every objectionable web page worldwide. Interviewed by *Wired* magazine, the computer engineer overseeing the *fanghuo qiang* bluntly explained his working policy: "You make a problem for us, and we'll make a law for you."

In many countries such a firewall might have stifled development, but most Chinese weren't interested in making problems. They were attracted to the internet's dazzling potential, as advertised on billboards that encouraged them to "join the internet club, meet today's successful people, experience the spirit of the age, drink deep of the cup of leisure." Those who could afford a connection, which cost approximately half the monthly salary of a recent college graduate, casually referred to the *fanghuo qiang* as the *wangguan*, calmly evoking the many *guan* (passes) of the Great Wall as natural features of China's *wan wei wang* (ten-thousand-dimensional web). Noting this local slang, the June 1997 issue of *Wired* presented the newfangled Chinese internet to Western audiences with a decidedly ironic spin, introducing the *Great Firewall of China* into the lexicon.

Since the 1990s the Great Firewall has grown stronger, much as the Ming emperors fortified the physical barriers of former dynasties by supplementing mud with masonry. The firewall has also been made more effective, much as the Ming improved surveillance and communication by adding watch towers.* Impervious to irony, the Chinese government has dubbed this eight-year, $700 million venture the Golden Shield Project. Like the original *wangguan*, the Golden Shield can block access to websites deemed "unhealthy" by the government, such as the BBC News and Amnesty International. Unlike the original *wangguan*, the Golden Shield can also filter out search engine results containing "spiritual pollution." Search for the words *Dalai Llama* or *democracy* in Mandarin, and you're likely to come up with no matches. Moreover, the Golden Shield can automatically punish untoward curiosity with "time-outs." Try to reach a forbidden URL and internet access will be suspended for thirty seconds. Try again and the suspension may last three minutes. Try a few more times and you might be put offline for hours—no explanation given—while the Public Security Bureau is alerted.

The Western press has reported widely on these draconian measures, almost always under the Great Firewall rubric. The media has rightly condemned the Chinese government for oppressing the population and unjustly punishing dissidents, such as the intellectuals who distributed the Charter 08 petition demanding democratic reform. (The Public Security Bureau responded by shutting down nearly a thousand websites and 250 blogs, ostensibly to "purify" the web of pornography, which together with "state secrets" serves as a euphemism for anything that exposes the power structure.) In this strict sense *Great Firewall* is a misnomer: the Great Wall of China was a defensive barrier against invasion, not a prison.

---

* The watch towers were the ethernet of their era. Signals about enemy positions could be relayed along the barrier, using smoke by day and fire by night, at an astounding speed of twenty-six miles per hour, considerably faster than a man on horseback.

However, the Great Wall of China as a metaphor for the *fanghuo qiang* has persisted not in terms of function, but rather with reference to the wall's ultimate demise, in 1644, when foreign invaders unseated the Ming after 276 years in power. "As every Chinese school kid knows," observed *Wired* in 1997, "the original Great Wall failed in its basic mission." Ten years later almost exactly the same argument was made in a second *Wired* feature: "No shield, golden or otherwise, can protect [the government] from the public. China's leaders should know this. Their predecessors built the Great Wall of China to keep out Mongol invaders. It proved as useful as every other fixed fortification in history, and the Mongols still invaded Beijing and overthrew the political elite." Like numerous other Western sources, from the *New York Times* to the BBC, *Wired* supported this position by detailing the ways Chinese citizens routinely penetrate the wall using encryption and proxy servers. The methods work, and for a simple reason: businesses also depend on them, and the Chinese government depends on business.

So why does the Great Firewall endure? The primary reason is that the metaphor is more accurate than intended by clever Western commentators. The greatest impact of the *wangguan* within China, like the *guan* of the Ming, is as a protective fortification against foreign conquest. Only in this case the prospective invaders aren't Tatar, Manchu, and Mongol. They're eBay, Twitter, and Google. Through measures ranging from rerouting to blockage, the Great Firewall has screened out these foreign companies effectively enough for Chinese equivalents to dominate the domestic market. Most Chinese go to Alibaba instead of eBay, Tuscent instead of Twitter, and Baidu instead of Google. And given that China surpassed the United States as the largest online population scarcely a decade after the Potemkin Year of the Internet, the Chinese domestic market is large enough to render the foreign market essentially irrelevant.

In fact the Great Firewall may achieve what the Great Wall of China could not, for the Chinese have indeed learned from the past. The Great Firewall is permeable. It resembles less a fortress than a speed bump. Foreign search engines are seldom blocked outright, and with a little subversion, generally over-looked by the government, a citizen can even access the banned BBC World Service. But the minor inconvenience of doing so extracts its toll in terms of raw statistics. Most Chinese, like most Americans and British, are lazy. Certainly they want to "meet today's successful people, experience the spirit of the age, drink deep of the cup of leisure," but only with the least possible effort. The real barrier is the average user's apathy.

The Great Firewall is a psychological wall. Its strength as a fortification is the strength of a prison built from within.

Flog

A flack blog.

In the fall of 2006 a typical American couple named Jim and Laura drove an RV from Nevada to Georgia, blogging about their encounters with Wal-Mart employees and customers. Because Wal-Mart was one of their favorite stores, they found plenty to praise, including the morale of clerks and corporate health care benefits. Yet there was another reason their excursion reflected so well on the notoriously ruthless company: the trip was financed, and their blogging paid for, by Wal-Mart's public relations agency.

Less besotted by big box stores than were Laura and Jim, other bloggers soon began taunting them, publicly questioning whether WalmartingAcrossAmerica.com was a sham. The backlash threatened to go viral. The site was hastily dismantled, and the PR firm brusquely apologized. Yet the incident was immortalized on the strength of a word that perfectly embodied the flacks' marketing folly. Walmarting Across America became the first big flog.

Flog is not an unusual coinage for the web, where words are routinely mashed up to accommodate intersecting ideas and high-speed typing. Films combining porno and gore are sometimes dubbed *gorno*, and the proliferation of girdles for men has begotten the *mirdle*. Even *flog* has had several other incarnations, including abbreviations for *family blog*, *food blog*, *photo blog*, and *For the love of God*. What distinguishes the current example is the cunning play on words, the sly (if less than subtle) reference to flogging, old slang for selling goods of dubious merit, derived from cant for *flagellation*. Used in reference to flack blogs such as Walmarting Across America, *flog* sounds like what it is: a term for PR chicanery. *Flog* has much to recommend it linguistically, not least its appropriation of *blog*, one of the most successful neologisms in Internet history. Yet despite its mix of pedigree and wit, *flog* is well on its way to oblivion.

Flog has foundered for many of the reasons that *blog* has flourished, and their apparent similarities reveal their real differences. Both words originated as contractions, yet *blog* was almost as arbitrary as *flog* was deliberate. "For what it's worth, I've decided to pronounce the word 'weblog' as wee'- blog. Or 'blog' for short," announced an information architect named Peter Merholz on his homepage in April or May 1999. At the time his aside wasn't worth much, since the number of people keeping weblogs, a two-year-old marriage of the words *web* and *log*, was perhaps in the thousands, and the number of people reading his weblog was far fewer than that. However, one reader happened to be Merholz's friend Meg Hourihan, who was developing a service to make weblog production easy for amateurs and needed a distinguishing trademark. She and her partners called their company Blogger and began touting the virtues of blogging.

Blogging took off partly because services such as Blogger made it accessible, partly because petty narcissism became a trendy corollary to online anonymity, and partly because the trend had a fresh name, a shibboleth for a solipsistic subculture: you'd never guess what a blog was unless you'd read one. Merholz later claimed that he considered *blog* to be "roughly onomatopoeic of vomiting," admitting that blogs "tend to be a kind of information upchucking." He had a point, though the reference presumably had as little to do with the widespread adoption of the term as with the initial coinage. From the start the connotations of blogging were positive, actively promoted by participants who enthusiastically identified themselves with the term *blogger*. Word and deed were mutually reinforcing, multiplying exponentially in a positive feedback cycle until, at last count, there were 112 million blogs, and 100,000 new ones launching every day.

Flogging, on the contrary, subsists in a negative feedback cycle. The more the word gets used, the less likely potential floggers are to provide future opportunities for use. Either they'll flog more surreptitiously, or they'll resort to new ploys. Following the Wal-Mart debacle only two other major flogging incidents have been documented, both in 2006. McDonald's engineered a couple of flack blogs ostensibly created by customers obsessed with a Monopoly cross-promotion, and Sony launched "All I want for Xmas is a PSP," which was presented as an attempt by a teenager to lobby his friend's parents to buy his buddy a gaming console. But real bloggers recognize their own kind. All three flogs were rapidly exposed, ridiculed, and withdrawn.

The guises that viral marketing can take are practically unlimited. (Sony has also hired mercenary graffiti artists to chalk up city streets with pictures of kids playing with PSPs.) Flogging is as parasitic as the rest, and has proven less hardy than most. But if the word disappears with the practice, it will have served its purpose. The coinage was clever enough to have induced its own obsolescence.

CROWDSOURCING

Outsourcing to the masses.

On the day that the June 2006 issue of *Wired* magazine was released, the publication's technology director searched the web for the word *crowdsourcing*, the subject of an article by contributing writer Jeff Howe. He took a screenshot of what he found, a total of three brief mentions, and forwarded it to the author, advising that Howe save it as a "historical document." Howe didn't have to wait long to see history in action. Within nine days Google was returning 182,000 hits. Nor was it a fleeting fad. Three years later the number had multiplied to 1,620,000, with regular appearances in the mainstream media, from the *Washington Post* to Fox News, where *crowdsourcing* was averaging two hundred new mentions each month.

There's a simple explanation for the neologism's success. Howe had detected a trend and given it a word. The backstory, which Howe posted on his personal blog, crowdsourcing.com, supports this notion:

In January Wired asked me to give a sort of "reporter's notebook" style presentation to some executives. I had recently been looking into common threads behind the ways advertising agencies, TV networks and newspapers were leveraging user-generated content, and picked that for my topic. Later that day I called my editor at Wired, Mark Robinson, and told him I thought there was a broader story that other journalists were missing, ie, that users weren't just making dumb-pet-trick movies, but were poised to contribute in significant and measurable ways in a disparate array of industries.

In what Howe characterizes as "a fit of back-and-forth wordplay," he and Robinson came up with a term that riffed off the title of a business book popular at the time, James Surowiecki's *The Wisdom of Crowds*, while also suggesting open-source software and corporate outsourcing. When "The Rise of Crowdsourcing" was finally published six months later, the last of these three roots was explicitly evoked in the teaser: "Remember outsourcing? Sending jobs to India and China is so 2003. The new pool of cheap labor: everyday people using their spare cycles to create content, solve problems, even do corporate R & D."

The article set out to prove that this was happening. To make his case Howe rounded up four prime examples. There was a VH1 television show that broadcast viewers' homemade videos, and a stock photography agency, iStockphoto, that allowed amateurs to post their pictures, getting paid a nominal fee each time an image was used commercially. There was a company called InnoCentive through which large corporations such as Procter & Gamble could publicly solicit technical solutions to manufacturing problems, such as how to get fluoride powder into a toothpaste tube. And finally there was Mechanical Turk, an Amazon.com service that let companies contract out menial work, such as sorting through digital photographs, to anyone with some spare time. Taken together these represented a spectrum of activity so broad that other instances of it were inevitable, enough to fill a 320-page book published by Howe two years later.

More important for the word's endurance, the breadth invited adoption of the term by others. Many have since seized the opportunity. Over a typical three-day period in 2009, for instance, the *Guardian* reported that Utrecht University planned to crowdsource its next advertising campaign, the *Examiner* covered the debut of an auto manufacturer crowdsourcing vehicle design, and the *New York Times* announced that it would crowdsource a special report about what locals are reading by observing the books in people's hands on various subway lines. In those same seventy-two hours Reuters mentioned *crowdsourcing* in three separate wire stories, and five articles used *crowdsourcing* just with respect to Wikipedia.

Not surprisingly Wikipedia's own entry on the topic has become a veritable crowdsourcing directory. The page lists companies that are crowdsourcing software debugging, graphic design, transit planning, and gold prospecting. Moreover volunteers are being crowdsourced to spot new galaxies, monitor illegal U.S. border crossings, and investigate salary abuses by British MPs. Participation can be incredibly lucrative, as in the case of the $1 million Netflix Prize, awarded for improving the company's video recommendation algorithm. Or involvement can be purely entertaining, as in the case of Foldit, a game that crowdsources protein-folding problems by formulating them as brainteasers. There's an average of one update per day to the Wikipedia entry. Which is to say that Wikipedia is effectively crowdsourcing the definition of crowdsourcing.

That's not what Wikipedia's cofounder (and de facto editor in chief) Jimmy Wales would have people believe. In one typical tirade Wales told the *San Francisco Chronicle* that he considers crowdsourcing "a vile, vile way of looking at that world. This idea that a good business model is to get the public to do your work for free—that's just crazy. It disrespects the people."

Though arguably disingenuous in his remarks and hypocritical in his beliefs, Wales is hardly alone in his criticism of crowdsourcing. Some of the most vehement protest has come from the graphic design community, in opposition to crowdsourcing portals such as crowdSpring, a company that provides businesses with a platform for farming out creative work to anyone with a copy of Adobe Illustrator and a willingness to work on spec. Designers understandably view crowdsourcing as a threat to their livelihoods, enticing potential clients to contract hobbyists, depressing prices and eroding standards. Other creative professionals do too, including freelance photographers and writers.

Befitting the broadness of the term, there have also been broader critiques. For instance, at the 140 Characters Conference, a two-day forum on the social effects of Twitter, the freelance rock critic Christopher R. Weingarten asserted, "Crowdsourcing killed indie rock . . . 'cause crowds have terrible taste." And policy analyst Pete Peterson cautioned that the "government needs smart-sourcing, not crowdsourcing" in an essay about online opinion polling on the wonky political blog techPresident.com.

Public policy and indie rock seldom cross paths, and one might be excused for asking how they relate to the graphic design industry, let alone Wikipedia. To encompass all of these, the definition of *crowdsourcing* has become so amorphous as to defy even Howe's best efforts. (On his blog he defines it as "the act of taking a job traditionally performed by a designated agent (usually an employee) and outsourcing it to an undefined, generally large group of people in the form of an open call," a fair description that clearly doesn't cover Weingarten's commentary on public taste or Peterson's on representative democracy.) The truth is that *crowdsourcing* can no longer be delineated with a set of necessary and sufficient conditions. Cases of crowdsourcing have rather come to possess what Ludwig Wittgenstein referred to as a *family resemblance*.

Wittgenstein introduced the idea of family resemblances in *Philosophical Investigations*, discussing the problem of defining *language*. "Instead of producing something common to all that we call language," he wrote, "I am saying that these phenomena have no one thing in common which makes us use the same word for all,—but that they are *related* to one another in many different ways." Wittgenstein observed that many words have this characteristic, and that considering the multiplicity of uses reveals "a complicated network of similarities overlapping and criss-crossing: sometimes overall similarities, sometimes similarities of detail." The meaning of *crowdsourcing* is to be found in the network of similarities between representative democracy, public taste, Wikipedia, et cetera.

Conversely the word *crowdsourcing* reveals hidden ways in which these phenomena are interconnected. That is the real strength of the word, and of any term that overgrows the trend it was coined to define. (*Open source* is another good example, having come to encapsulate copyleft values well beyond collaborative software design.) Debates about crowdsourcing reflect genuine dilemmas in contemporary society. There's no obvious balance to be found between professional integrity and freelance opportunity in the feud between graphic designers and crowdSpring, or between expert judgment and majority opinion in the valuation of indie music. However, these dilemmas can inform each other, and because they're so different the issues at stake become more explicit when considered together. Embodied in the collective discussion is the wisdom of crowdsourcing.

PART IV

PROMOTION

Reports of alien spacecraft flying over the southwestern American desert suffered a serious setback in the late 1940s, after descriptions of their dish-like shape inspired a catchy nickname: skeptics dubbed them *flying saucers*. To close the credibility gap, believers adopted the bureaucratic-sounding term *unidentified flying object*, and further distinguished it with an acronym. UFOs became a topic of mainstream conversation on the basis of a simple change in phrasing.

Coined to promote a cause, *UFO* served its purpose by entering into the language practically undetected, appearing to be just another piece of military jargon. Talk of UFOs sounded familiar, like *PX*, and legitimate, like *NATO*.

Many words today are likewise circulated for purposes of promotion without bearing the official status of a brand name, let alone the legal protection of a trademark. Some of these, such as *tweet*, work to the advantage of a specific company. Others, such as *conficker* and *steampunk* and *lifehacker*, serve an entire industry or subculture. These words have caught on by slipping into the public domain. Their partisans gain influence by losing control. But there are no guarantees. Language has an honest streak. For flying saucer enthusiasts want of evidence eventually turned terms such as *ufology* into a fresh lexicon of ridicule.

CONFICKER

Alias for one of the most infectious computer worms of all time.

Some called it *Downadup*. Others preferred *Kido*. As soon as the Conficker worm started spreading over the internet on November 20, 2008,[*] security firms agreed that they faced a pandemic and immediately began to cooperate on containing it, but they couldn't reach a consensus about what to name it.

The worm continued to propagate unabated, infecting an estimated 15 million computers, including systems in the German military and British Parliament. By the following April the media had reported on the worm so extensively that *PC World* compared its notoriety to the celebrity of Paris Hilton, dubbing Conficker "the world's most famous piece of malware." Nevertheless several security companies still insisted on calling it Downadup. And though nobody was any closer to eliminating

---

[*] While the terms *worm* and *virus* are sometimes used interchangeably—and worms and viruses do similar jobs for a hacker—a worm is technically a piece of malware that runs on its own, whereas a virus works by commandeering software already on a computer. In this respect they are roughly analogous to their biological namesakes, the parasitic worms and viruses found in nature.

it, the worm had acquired several more identifiers, official codes such as TA08–297A, VU827267, and CVE-2008–4250, seldom referenced by anyone.

Conficker will always have conflicting monikers. Most malware does. Despite periodic attempts to standardize the naming process, no system has ever become ubiquitous. Even the first, codified in 1991 when a mere thousand viruses were in circulation, was ignored as often as it was followed by the half-dozen computer firms then in the nascent security business. And no wonder. Mandating the form *Family_Name.Group_Name.Major_Variant.Minor_Variant[:Modifier]*, it made Linnaean binomials seem, comparatively, to roll off the tongue. A 1999 revision only bloated the nomenclature, requiring that *virus platform* and *malware type* also be specified, while ignoring a much deeper problem: even researchers who followed the formula seldom agreed on what to call the groups and families. Each new worm or virus averaged four totally incompatible appellations—generally unpronounceable strings of letters—and every month the number of worms and viruses in the wild increased by thousands.

Hackers spread much of the malware by email. You'd get a virus embedded in the attachment to a message ostensibly coming from a personal contact. If you opened the attachment the virus would take over your computer's email software, automatically forwarding itself, now in your name, to every address on your own contact list, renewing the whole infection process. The technique, which left infected computers under the hacker's control unless the virus was removed, required more than just good software programming. To convince people to open the attachment also took clever social engineering. Hackers gradually learned how to pique recipients' interest without arousing their suspicion. And one of the greatest successes ever was an attachment that began circulating in 2001, simply labeled AnnaKournikova.jpg.vbs.

Anna Kournikova was a tennis player, at the time ranked in the top ten worldwide, but better known for her sports-bra modeling and her regular appearance on *People* magazine's 50 Most Beautiful People. In other words, a picture of her promised to be sexy but not scandalous, just the sort of thing a typical guy might forward or download without a moment's thought. They did so in droves, spreading the virus so quickly that the BBC reported "paralysed e-mail servers around the world" two days after it was released. Like most news reports, the BBC article mentioned in passing some of the malware's official names, VBS and SST, but dubbed it the "Kournikova computer virus" in the headline and illustrated it with a picture of the tennis player. The media knew how to package a story that would attract readers. After all, they'd helped make the middling athlete a star, much more famous than players who routinely beat her.

Soon everyone outside the security industry was calling the virus Kournikova or AnnaKournikova, and though the malware never caused lasting damage the name gained it iconic status (including a mention on the TV sitcom *Friends*). A lot of antivirus software was bought on that basis, and—security firms taking notice—a cottage industry in catchy virus naming was born.

Firms vied to come up with names that would lend malware charisma, as *AnnaKournikova* had done, making viruses and worms into media starlets. It was a peculiar business, since the antivirus software vendors were effectively marketing themselves by marketing their adversaries. The chaos of naming viruses and worms morphed into a sort of branding war, in which a company that came up with a popular moniker could bask in the reflected glory of the malware.

Some of the most mediagenic namings, such as KamaSutra, were poached directly from the infected emails' alluring subject lines, suggesting just how closely the interests of hackers and their pursuers may be intertwined. But the greatest popular success before Conficker came in 2004, when a researcher for McAfee Avert, Craig Schmugar, started taking apart a virulent new worm, finding a line of code reading "mydom," an abbreviation of "my domain." Drolly doubling the *o* he dubbed the malware MyDoom and watched it become a household name. "It was evident early on that this would be very big," he told *Newsweek* days after the worm began circulating, by which time it had infected 500,000 machines. "I thought having 'doom' in the name would be appropriate." Alternate handles such as Novarg and Shimgapi didn't stand a chance, and even a later attempt to snag attention for a variant by calling it Bofra went nowhere.

When the *Newsweek* article ran, the true purpose of MyDoom was coming into view. In addition to spreading the worm by email and through file-sharing services, infected "zombie" computers were being made to attack the websites of Microsoft and the SCO Group, both software companies loathed by anticorporate hackers. It was an old technique, dating back to the 1980s: overload company servers by having hordes of zombies access them simultaneously. SCO's site was shut down briefly in this way, but the hackers gained far more notoriety, and attention for their cause, courtesy of McAfee Avert's compelling branding of the worm. Routine denial-of-service attacks made headline news because they were committed by a virus named MyDoom.

Conficker's purpose is more murky, and its celebrity is less obvious. Like *MyDoom*, *Conficker* is borrowed from the malware code, in this case rearranging the domain trafficconverter. biz. Unlike MyDoom (or KamaSutra or AnnaKournikova), the name is not overtly evocative. What seems to have given it such resonance is a bogus folk etymology, popularized on Wikipedia:

"The origin of the name Conficker is thought to be a port-manteau of the English term 'configure' and the German word Ficker, which means 'fucker.' " While *Downadup* may have the virtue of sounding like a hip-hop alias (à la Eminem), *Conficker* is the first malware name to resort, ostensibly, to fornication.

Still, that's not the reason for the comparisons to Paris Hilton. As *PC World* explained the resemblance, both the malware and the heiress are "famous solely for being famous. Neither has actually ever done anything of note." These words were published on April 1, 2009, the day that virus researchers examining the code believed the true goal of Conficker would be revealed, since the software was programmed to take new orders on April Fool's. Expectations were high, with experts claiming it could bring down the internet. When it didn't, there was a slight tinge of disappointment.

Six days later Conficker quietly began doing something other than spreading: some of the computers started sending out spam, selling, of all things, fake antivirus software. So apparently Conficker was not the work of guerrillas seeking to sabotage the German military or the British Parliament, let alone the whole internet. If the purpose of Conficker was illicit commerce, not public protest, the attention brought by its popular moniker must have been unwelcome. Simply by naming the worm security companies may have undermined it in much the same way that MyDoom's effect was boosted.

Yet hackers maintain a crucial advantage over their pursuers. Careful scrutiny of malware may reveal how it works, but no amount of analysis can betray a worm's purpose. Some researchers first thought that MyDoom was made to send spam, and that may have been the case. Then it found fame. The objective might have changed with the fate of its name.

STEAMPUNK

Sci-fi set in the steam age.

"I wish to God these calculations had been executed by steam," exclaimed the British polymath Charles Babbage to his colleague John Herschel one day in 1821, as they worked together to correct a batch of mathematical tables riddled with errors. With that outburst, according to his memoirs, Babbage envisioned the first computer. The machine he conceived was colossal, a cogwheel behemoth comprising twenty-five thousand parts, planned to measure seven feet long and to weigh fifteen tons. The British government invested £17,500 in it—the cost of twenty-two new locomotives—yet after eleven years of hard labor Babbage's unfinished difference engine was abandoned.

But what if construction had succeeded? That's the question sci-fi writers William Gibson and Bruce Sterling asked a century and a half later, their answer serving as the premise of *The Difference Engine*, a novel in which the information age overtakes Victorian England. As a work of speculative fiction the book was a deep meditation on the interdependence of technology and society, destined to have an intellectual impact nearly as significant as Gibson's breakthrough *Neuromancer*, in which he introduced the idea of cyberspace. Also like *Neuromancer*, arguably the first cyberpunk novel, *The Difference Engine* was to spawn a vast subculture.

Steampunk, as the cult was dubbed, was actually named several years before *The Difference Engine* was published, in a 1987 letter to the genre magazine *Locus*, penned by the sci-fi writer K. W. Jeter. "Personally, I think Victorian fantasies are going to be the next big thing, as long as we can come up with a fitting collective term," he wrote. "Something based on the appropriate technology of that era; like 'steampunks,' perhaps." James P. Blaylock, another writer of these "Victorian fantasies," seconded Jeter's suggestion in the following issue, and the subgenre was sufficiently established by the time *The Difference Engine* was published in 1990 that the *Locus* editors decreed it "*not* steampunk, because it is a work of hard sf."

In other words, the book responsible for popularizing steampunk wasn't considered steampunk by those who introduced the term, a paradox that has repeated itself ever since: with each new level of recognition, old definitions have been overwritten. By 2008, the year steampunk definitively went mainstream, manifesting in rock concerts and fashion shows and scoring coverage in venues from MTV to the *New York Times*, the number of elder steampunkers dismissive of what steampunk had become could practically be considered a cult in its own right. "Steampunk's in the *New York Times*," blogged Bruce Sterling the day it hit page 1 of the Style section. "So the 'death of steampunk' must be near." Of course, as his quotation marks suggest, he recognized that this latest death was just the newest resurrection.

In the *Times* story another steampunk veteran attempted to explain this mechanism. "Part of the reason it seems so popular is the very difficulty of pinning down what it is," said Sean Slattery, known by the steampunk pseudonym Hieronymus Isambard "Jake" Von Slatt. "That's a marketer's dream." Slattery belongs to the first generation to wrangle steampunk from the written page, modifying present-day computers to resemble Victorian machines by recasing the electronics in wood and brass and replacing buttons and switches with levers and cranks. Many steampunk "contraptors" (as they call themselves) extend the game by crafting apocryphal nineteenth-century inventions from antique or custom-fabricated engine parts, giving them fanciful names, such as Dr. Grordbort's Infallible Aether Oscillator and Lord Featherstone's Clockwork Arm. Some of these objects have been exhibited as sculpture at events such as Burning Man, though they're too superficially decorative to be taken seriously in artistic or scientific circles. (One exception is the London Science Museum's decade-long construction of a working difference engine, following Babbage's original plans, an epic feat of intellectual history that, tellingly, practically no one has dubbed *steampunk*.)

Art and science, in any case, were just way stations. Far dreamier to marketers was the crossover from philosophical toys to retail merchandise. Steampunk attracted the attention of MTV and the *Times* Style section when trendsetters such as the goth band Abney Park started outfitting themselves in steampunk clothing and accessories and presented themselves as "air pirates," templates for teen imitation.

By then *steampunk* was a historical hodgepodge, applied to pretty much anything made to appear invented between 1650 and 1950, while the literary genre, expanded to include role-playing games, had splintered into nearly a dozen subcategories, including dieselpunk, biopunk, and clockpunk, identifiable only to their respective fan bases. The terminology's paradoxical development and diminishment were finally complete.

We might think of what happened as a linguistic tragedy of the commons. Overexploited at the expense of cooperation, the term has been reduced to meaninglessness, the conceptual remains protectively renamed by noncommunicating factions. Alternately we might cast the linguistic misadventure in more positive terms and remark that the language has accurately reflected the ideas at every stage. Both peaked with the publication of *The Difference Engine*, started to drift with the trifling fantasy machines of the steampunk contraptors, and descended into vapidness with the neoretro posturing of the airship pirates.

Tweet

To send a text message on Twitter.

The big news in the Twitterverse on October 19, 2009, was the sighting of the pentagigatweet. Sent by an out-of-work dot-com executive named Robin Sloan, the six-character text message, a bit of banter between friends, garnered more attention than the war in Afghanistan or the swine flu pandemic. "Oh lord," it read.*

The message was sent at 10:28 a.m. PST. By 3:47 p.m. a CNET news story proclaimed, "Twitter hits 5 billion tweets," quoting Sloan's two-word contribution to telecommunications history, and noting that he'd geekily dubbed it the *pentagigatweet*. The following day newspapers around the world, from the *Telegraph* in England to *Il Messagero* in Italy, had picked up the story, yet the most extensive coverage was on Twitter itself,

---

* Sloan's tweet was inadvertently evocative of the first message ever sent electronically, "What hath God wrought," telegraphed to Baltimore from the U.S. Capitol by Samuel F. B. Morse on May 24, 1844—though in keeping with Twitter style, "Oh lord" was considerably more colloquial.

where nearly 30 percent of the estimated 25 million daily messages referenced the benchmark. The numbers were impressive. But more remarkable than the level of popularity achieved in the mere thirty-eight months since the microblogging service launched in 2006 was the degree to which those who used it felt responsible for building it. The megatweeting greeting the pentagigatweet was a sort of collective, networked navel-gazing. In the days following the five billionth text message Twitter was atwitter with self-congratulation.

That sense of personal investment, essential to Twitter's growth, was entirely by design. As Jack Dorsey explained in an interview with the *Los Angeles Times* about the company he cofounded, "The concept is so simple and so open-ended that people can make of it whatever they wish." Dorsey based the service on his experience writing dispatch software and his insight that the best way to observe a city in real time was to monitor the dispatches coming from couriers and taxis and ambulances. Twitter was created to put that experience in the hands of ordinary citizens, literally, by asking people to periodically send in text messages by mobile phone answering the question "What are you doing?" All participants would be able to follow the stream of responses. In other words, Twitter was formulated as a sort of relay, utterly dependent on the public for content. And to attract a suitably diverse demographic, Dorsey and his partners knew that the service, blandly called *Status* during the development phase, would need an image more enchanting than vehicle dispatch.

They started the naming process with the sensation of a cell phone vibrating, which brought to mind the word *twitch*. Since that didn't "bring up the right imagery," as Dorsey later recalled in his *Los Angeles Times* interview, "[they] looked in the dictionary for words around it, and [they] came across the word 'twitter,' and it was just perfect. The definition was 'a short burst

of inconsequential information,' and 'chirps from birds.' And
that's exactly what the product was." The bird metaphor was
especially appealing to Dorsey: "Bird chirps sound meaningless
to us, but meaning is applied by other birds. The same is true of
Twitter: a lot of messages can be seen as completely useless and
meaningless, but it's entirely dependent on the recipient."
What's more, birds are adorable, especially when drawn in cartoon caricature, as in the case of the critter on the Twitter website. If *status* sounded coldly diagnostic and *twitch* brought to
mind a case of Tourette's, *twitter* conveyed cuteness.* It was the
sort of application you might adopt like a pet, developing a
sense of attachment, a relationship.

And that's precisely what has happened. The Twitter community is fanatical. Even the Fail Whale—a cartoon depicting
a whale being hoisted out of the water by a flock of birds,
posted on Twitter whenever the server is overloaded—has a
Facebook fan club and has been reproduced on T-shirts peddled at Fail Whale parties featuring Fail Whale cocktails (vodka,
Cointreau, and blue curaçao). Licensed by Twitter for several
dollars through iStockphoto, the Fail Whale was never intended
to go primetime. It wasn't even supposed to have a name. Twitter users came up with one, just as they've come up with an
entire language of twitterspeak, as catalogued on websites such
as Twittionary and Twictionary. *Twitizens* follow *twaffic* to find
out what's *twendy* and *twittastic*.† Yet even more than *twitter*, the

---

*   One measure of the word's intrinsic charm is that it was previously picked up by
    Disney. The 1942 movie *Bambi* featured the song "Ev'rything Is Twitterpated in the
    Spring." *Twitterpated* is now shorthand on Twitter for getting addled by too many
    messages.
†   Most twitterspeak, in keeping with the adorable packaging of Twitter, has the twee
    quality of infantile children's literature. One can't help but respond to the onslaught
    by quoting Dorothy Parker's legendary *New Yorker* review, under the pseudonym
    Constant Reader, of a 1928 sequel to *Winnie the Pooh*: "And it is that word 'hunny,'
    my darlings, that marks the first place in *The House at Pooh Corner* at which Tonstant
    Weader fwowed up."

word that has captivated users is *tweet*, which, in keeping with Dorsey's birdie imagery, evolved as slang in the Twitter community for a microblog update. *Tweetahaulic tweeple* meet their *tweethearts* at *tweet-ups*.[*] It is in this linguistic tradition that Robin Sloan named his pentagigatweet. And it is against this cultural backdrop that Twitter has come closest to alienating its tweeps.

The furor began on July 1, 2009, when the industry blog TechCrunch ran an article headlined "Twitter Grows 'Uncomfortable' with the Use of the Word Tweet in Applications." The story described a quibble between Twitter and another microblogging application developer, in which the former advised the latter that "Twitter, Inc is uncomfortable with the use of the word Tweet (our trademark) and the similarity in your UI and our own." The trademark on *Tweet* came as a surprise to TechCrunch and, as it happened, also to the U.S. Patent and Trademark Office, which had only recently received a trademark application and had not yet begun reviewing it. But most of all, twitizens were shocked, and many were appalled. The TechCrunch article drew 251 comments. One user grumbled, "They better be planning to piss off alot of people." "Getting power trippy much? meh," another scoffed. Several more posted dictionary definitions of *tweet*, often with remarks such as "Put a copyright on a dictionary word?" and pronouncements that Twitter was moving to lock up the English language.

Of course all of this was overstated, and most was inaccurate. Nonetheless, sensing a spirit of revolt, cofounder Biz Stone felt compelled to respond the same day on the company blog, "We have applied to trademark Tweet because it is clearly attached to Twitter from a brand perspective but we have no intention of 'going after' the wonderful applications and services that use

---

[*] Ibid.

the word in their name when associated with Twitter. In fact, we encourage the use of the word Tweet. However, if we come across a confusing or damaging project, the recourse to act responsibly to protect both users and our brand is important." His post did not exactly quell the controversy. More articles appeared, dozens of them, some sensibly pointing out how much Twitter benefited in functionality from third-party applications such as Tweetie, TweetDeck, BackTweets, and Tweet-Meme. Much like the social model of Twitter, the business model had been to let outside developers make of it whatever they wished; Twitter's brand identity was built primarily by companies that, now that Twitter was an internet powerhouse, might perversely become liable for trademark infringement. Yet the deeper indignation came from ordinary users who had no legal stake in whether *tweet* was trademarked, but felt emotionally violated when reminded that Twitter was not communal property and that there was a corporate entity behind the cute birdie interface.

On one level Twitter was encountering a classic dilemma of branding. In order for a business to gain market share a brand needs to enter the vernacular, but once it's there a company must protect it from becoming generic. That's what Xerox has done, making *xeroxing* synonymous with *copying*, only to insist that copies made on other company's machines not be called *xeroxes*. Trademark law codifies this by insisting that companies actively protect their trademarks or lose their legal claim to them—through a process informally known as genericide—as has happened with former brand names from *escalator* to *yo-yo* to *aspirin*.

On a deeper level Twitter was coming up against the increasingly troubled relationship between trademarks and new media, where brands are not made by companies for consumers, but where consumers and third parties (often one and the same) collectively give brands their identity. Like copyright law, trademark law was formulated in a pre-internet society. Products such as Kleenex and Jell-O were developed by companies that brought them to market for consumers to purchase, a straightforwardly linear process that trademark law supported by allowing manufacturers to uniquely identify their merchandise, establishing a reputation and preventing customer confusion. For this reason trade names were static and descriptive, legally designated proper adjectives. There was Kleenex tissue and Jell-O gelatin, and on the rare occasion that a brand grew so predominant that the name became shorthand for the product or even for what it did, as in the case of Xerox, the company went on a campaign to make people cease and desist. However, in the case of Twitter, predicated on the idea that consumers should make of it whatever they wish, the brand as proper adjective lacks description. Or rather the process of definition is active, worked out in conversation, where adjectives and nouns and verbs naturally grow out of each other. People tweet on Twitter by twittering tweets. That is the open-ended way such a company develops an identity, whether or not the Patent and Trademark Office condones it.

The effect of trademark law has been to turn companies such as Twitter into language police. Perhaps no corporation has assumed this role more forcefully than Google, which seems to have an entire department specializing in sending cease-and-desist letters to anyone who publicly uses the company name as a verb. Notices have been sent to lexicographers such as Paul McFedries, who innocently defined *to google* on his WordSpy website in 2001, and even to the *Washington Post*, which in 2006 had the audacity to report on the addition of *google* as a transitive verb to the *Merriam-Webster's Collegiate Dictionary*. Given that the alternative to such vigilance is genericide, Google's compulsion is an understandable business decision. The question is whether trademark law itself is still viable or has become an artifact of the past, as incoherent as "community standard" legal restrictions on pornography in a networked society.

A few companies are starting to eschew legal precedent. Most notably, and unexpectedly, Microsoft has taken an unrestrictive approach to branding the Bing search engine, with CEO Steve Ballmer telling the *New York Times* that the name *Bing* has the potential to "verb up."* But it's debatable whether Microsoft lawyers will really allow this to happen if the Bing name gains traction: permissiveness is often just a form of desperation.

What makes *tweet* notable is that, having verbed and nouned up on Twitter of its own accord, it conveys brand identity while behaving in the generic way of natural language. That might not be what Twitter wants or what the Patent and Trademark Office can abide,† but with the pentagigatweet passed and the decagigatweet pending it may signal the communal future of new media branding.

---

* Of course, verbing up is easier said than done and cannot be achieved by marketing alone. For instance, the "Do You Yahoo!?" advertising campaign hasn't tricked anybody into talking about yahooing. Ultimately the public will have its way with a brand name, as Microsoft is learning with the spread of the inadvertent use of Bing as an acronym for "But it's not Google."

† As of 2010 the trademark application is still under review.

LIFEHACKER

A master at optimizing everyday routines online and off.

At the Massachusetts Institute of Technology in the 1950s, some of the brightest students seldom attended classes. Instead they loitered around the Tech Model Railroad Club. The most brilliant were tapped to join the Signals and Power Committee, which rigged ever more elaborate systems of programmable track switches using nothing more sophisticated than telephone relays. Taking pride in their ad hoc wiring, which ignored all conventions of electrical engineering, they referred to themselves as *hackers*. Nothing was impossible for them; nothing was off limits. When MIT acquired its first computer in 1956, they infiltrated the control room, where they coerced the electronics to do tricks unintended by the manufacturer, using sine and cosine routines to code the first digital computer game.

As computers became more common, so did hacking. To program home computers in the 1970s no longer required the imaginative genius of the MIT Signals and Power Committee, and by the 1980s self-professed hackers ranged from professional software developers to adolescent cyberpunks. The latter proved considerably more interesting to the public, riveted by their ability to torment corporations and governments from their bedrooms. "A hacker—computer jargon for an electronic eavesdropper who by-passes computer security systems—yesterday penetrated a confidential British Telecom message system being demonstrated live on BBC-TV," reported the *Daily Telegraph* in a typical news story of 1983, the year that *War Games* hit the big screen. Old-school Signals and Power hackers fought valiantly against this linguistic turn, insisting that the young punks were *crackers* rather than *hackers*, but the media ignored the distinction, leading most new-school professionals to head off confusion by blandly presenting themselves as computer scientists or software engineers or information technology specialists. Aside from the occasional insider reference—ITs who troubleshoot security systems were sometimes known as "white-hat" hackers—the criminal connotation seemed permanent.

Then in 2004 a Silicon Valley technology writer named Danny O'Brien gave a forty-five-minute lecture at the O'Reilly Emerging Technology Conference titled "Life Hacks: Tech Secrets of Overprolific Alpha Geeks." Within a year technophiles worldwide, from computer scientists to iPhone addicts, were striving to become hackers again.

Lifehackers, to be specific. While almost none had attended O'Brien's lecture, or had even heard of him, for that matter, his idea that life could be "hacked" to make living more efficient proved as irresistible to tech geeks as Scientology was to Hollywood celebrities. Not that O'Brien had much in common with L. Ron Hubbard. In keeping with geek culture, his approach was emphatically open-source. He began by asking the overprolific alpha geeks he knew "the secrets of their desktops, their inboxes, and their schedules." What he discovered, he admitted outright, was "surprisingly dull." Befitting their decidedly un-Hollywood personalities, the nerds in his audience were seduced by this lack of mystique.

What were the boring things O'Brien picked up from his colleagues? His talk wasn't recorded, but the blogger Cory Doctorow took extensive notes, from which it's been reconstructed and studied as diligently as one of Aristotle's discourses. "Power-users don't trust complicated apps. . . . Synching is the new backup. . . . Geeks write scripts to take apart dull, repetitive tasks. They'll spend 10h writing a script that will save 11h—because writing scripts is interesting and doing dull stuff isn't." From these pointers O'Brien extrapolated some broader principles—"Good is good enough. . . . Automate the repetitive stuff. . . . Use simple tools to help you. . . . Just getting started is a good start"—which he posted on lifehacks.com, a blog he intended to be an ongoing lifehacking resource.

But by his own admission O'Brien wasn't himself an overprolific alpha geek. He procrastinated on his website until several others, including 43folders.com and lifehacker.com, had rendered it superfluous. And through these lifehacking blogs, especially the latter, lifehacking spread from the IT elite to the geeky mainstream.

Lifehacker.com was launched by the online conglomerate Gawker Media in January 2005, just under a year after O'Brien's talk, when a Gawker web programmer named Gina Trapani brought Doctorow's lecture notes to the attention of the site's publisher. Trapani was promptly appointed lead editor, responsible for posting some twenty lifehacking tips a day. Since readers could be advised that good was good enough and to automate the repetitive stuff only so many times a month, she soon began expanding the gamut of lifehacking. There were techniques to "synchronize bookmarks across multiple computers" and to "switch between Windows on your PC like they do on Macs," but there was also information on decidedly offline activities, such as how to avoid rush-hour traffic. Four years later one characteristic post promised the "top 10 tricks for creatively hiding your stuff," and another explained how to fry a perfect egg. ("Frying an egg is one of those things that seems like it should be so very simple, but the wrong combination of heat, time, and oil can make your potential perfect egg turn into a burnt and unappetizing mess.") Both of these tips were in the weekly top ten.

By then the website was defining a *lifehack* as "a clever shortcut or a lesser-known, faster way to get something done," and *Time* magazine was describing the site as "a 'Hints from Heloise' for the digital age," naming it one of the year's "25 Best Blogs." "It's all about doing more with what you already have," the *Time* editors enthused, though Trapani herself summed up the lifehacking ethos best in a *Wired* magazine interview with a rhetorical question: "Why waste human cycles?"

Cycles here refers to the clock cycles of a computer, loosely correlated to processing power. Speaking of human activity in terms of cycles is metaphorical, of course, yet the metaphor is telling: we're equated with machines, and our routines can be optimized in a manner equivalent to computer software. People are hackable. The move from synchronizing bookmarks across multiple computers to frying a perfect egg is not a stretch in the realm of lifehacking, but its logical conclusion.

On the other hand such activities bear scant resemblance to programming a computer game using sine and cosine routines, let alone model railroading with telephone relays. While the cult of efficiency has given hacking a positive spin in popular opinion, the lifehacking movement is no linguistic triumph for the Signals and Power Committee. It may be the ultimate philosophical defeat.

When the Signals and Power clique used the word *hack*, they were placing themselves in a long-standing MIT tradition of recreational cleverness. In the 1920s and 1930s, for instance, students repeatedly smuggled automobiles onto the roofs of campus buildings, a technical challenge undertaken for no good reason other than that the pranksters, often anonymous, had great fun doing it. The engineering taught in classrooms was straightforwardly purpose-driven; extracurricular hacking was the subversive opposite.

Old-school computer hackers have often struggled to characterize their activity, perhaps because it stands in such stark contrast to the pragmatism of technology and the practicality of American culture more broadly. GNU's architect Richard Stallman has described hacking as "playfully doing something difficult," citing Guillaume de Machaut's fourteenth-century musical palindrome "Ma Fin Est Mon Commencement" as one predigital example. The open-source Logo developer Brian Harvey more helpfully traces the hacking mentality back to Søren Kierkegaard, in contrast to Immanuel Kant, arguing that "a hacker is an aesthete."

It's hard to conceive of two ur-texts more distant than *Either/ Or* and "Hints from Heloise." Yet the distinction is revealing and helps to explain the earlier conflation of hackers and crackers, too easily attributed to media ignorance. No lifehacking alpha geek could be bothered to penetrate a confidential British Telecom message system being demonstrated live on BBC-TV. That belongs to the realm of playfully doing something difficult. A cyberpunk cracker is an aesthete as much as a hacker is one; moreover both are engaged in the same aesthetic project: to subvert routinely engineered progress. Lifehacking optimizes mankind as a technology. Hacking, whether undertaken by an adolescent cyberpunk or a Signals and Power Committeeman, tweaks technology to remind us of our humanity.

PART V

SLANG

One of the greatest novelties of the nineteenth century, the telephone presented a novel problem: If people too distant to see each other wished to communicate, how could they get one another's attention? Alexander Graham Bell proposed shouting *Ahoy!*, as was commonly done to hail ships at sea. Bell's estimable rival—and challenger to the telephone patents—counterproposed that people holler *Hello!*

Hello was not strictly new. Variants had been used for centuries as a generic call for attention, as in the line "Holla, approach" from Shakespeare's *Love's Labour's Lost*. This relatively unencumbered past—the fact that *hello* wasn't already characteristic slang of a specific trade—may have helped it to edge out *ahoy* on the phone lines (even as Bell edged out Edison in the courts) and also allowed *hello* to become shorthand for all telephony. (Operators, for instance, were originally known as *hello girls*.)

Online gaming, email, filesharing, and texting have each developed its own peculiar slang, as tightly coupled with those technologies as *hello* with the telephone. And like *hello*, which survived long after there was any technical need for it, broadening into a catchall greeting, these specialized vernaculars may be tomorrow's lingua franca.

Mashup

The merger of two or more independent sources to synthesize a
new stand-alone work.

The origin of the mashup is a matter of debate. According to
one theory, the phenomenon began in 2001 with the XFM
radio broadcast of the song "Stroke of Genius," a bootleg remix
by the deejay Freelance Hellraiser that incongruously set the
pop vocals of Christina Aguilera's "Genie in a Bottle" against
garage rock instrumentals from the Strokes' "Hard to Explain."
A competing hypothesis credits the culture-jamming Evolution Control Committee, which in 1993 satirically layered the
brutal rap lyrics of Public Enemy over swinging Latin arrange-
ments of Herb Alpert and the Tijuana Brass. Other theories cite
Club House's 1983 medley of Steely Dan's "Do It Again" and
Michael Jackson's "Billy Jean," Frank Zappa's '70s experiments
in xenochrony, King Tubby's '60s dub remixes, John Cage's '50s
compositions for a chorus of radios, and even the Renaissance
practice of quodlibet. Although some of these may have been
influential—and all are reminders of the role remixing has forever played in the creative process—this long tail of influences

scarcely anticipates the explosion of songs combining vocals from one source with instrumentals from another following the Freelance Hellraiser's XFM debut.

In a matter of months mashups numbered in the thousands, with juxtapositions including Missy Elliott vs. the Cure, Art Garfunkel vs. Watership Down, and Whitney Houston vs. Kraftwork. Evoking a wrestling match, *A vs. B* became the standard formula for citing sources, generally in parentheses following a title playing on names of the original songs. (For instance, "Smells Like Teen Booty" was a mashup of Nirvana's "Smells Like Teen Spirit" with "Bootylicious" by Destiny's Child.) The sounds of these remixes were as varied as the source materials, and the motivations were as disparate as the historical influences, with intended targets ranging from dance club entertainment to cultural critique. What these works shared, and have in common with the countless additional musical (and video) mashups that have since joined them, is the notion that culture is interactive, a feedback loop rather than a mail chute. Whether done in tribute or ridicule, or simply to create something beautiful, these songs mash up the standard distinction between consumer and producer.

For obvious reasons traditional producers have not welcomed this appropriation of power, any more than they embraced the older practices of sampling from or remixing tracks. As early as 2002, when mashups were still so new that the word was often hyphenated and set in quotation marks, Christina Aguilera's record label was fighting "Stroke of Genius" with a cease-and-desist order against XFM, and the British Phonographic Institute's Matt Phillips was telling the *Guardian*, "When you buy a CD you buy the rights to listen to it, not to change it in any way." Corporate types called mashups *bootlegs*, a term popular during the Prohibition era, which cast mashup artists in roguish terms they readily embraced. They also referred to their releases as *bootlegs* or *boots*, and the aura of subversively illicit activity became much of the appeal of even the most vapid pop mashups.

This is not the expected background for a term destined to become a buzzword in the corporate world. Yet within five years of the Freelance Hellraiser's "Stroke of Genius" *Business-Week* was reporting, "Already, companies such as E★Trade (ET), Siemens (SI), JDS Uniphase (JDSU), Pfizer (PFE), Glaxo-SmithKline (GSK), and Realogy (H) subsidiary Coldwell Banker Commercial are using enterprise mashups in some capacity." And in 2008 IBM announced "the industry's first complete mashup portfolio for business, empowering individuals to create situational applications, or mashups, and to help them do their jobs more effectively and meet the needs of the emerging real time enterprise."

There would seem to be no plausible connection between *mashup* as used by blue chip companies and *mashup* as used by freelance hellraisers, just as there is almost certainly no direct line from either of these back to the 1859 play by Dion Boucicault that the *Oxford English Dictionary* cites as the first occurrence of the word: "He don't understand; he speaks a mash up of Indian, French, and Mexican." In fact musical and enterprise mashups are closely related etymologically, and the terminology common to them reveals more about each than might be apparent if viewed in isolation.

The origin of enterprise mashups is almost universally traced back to HousingMaps.com, a service launched by the computer programmer Paul Rademacher in April 2005 while he was looking on Craigslist to buy a home in San Francisco. Frustrated by the difficulty of comparing the price and location of real estate, he hacked Google Maps to automatically overlay Craigslist offerings by neighborhood, and then put his homemade application online so that others could use it.

Among the first people to take notice was Peter Merholz, who'd coined the term *blog*—a foreshortening of *weblog*—on his website in early 1999. Merholz showed his understated knack for catchy language once again in his April 11, 2005, post: "One thing currently making its way around the Blogosphere is Paul Rademacher's Google Maps + Craigslist Housing mash-up, where you use the former to visualize the latter." Two days later an official Google blog picked up the word, saying, "Paul Rademacher's mashup of Craigslist and Google Maps blew our minds right off our shoulders," and by July HousingMaps.com and several similar hacks inspired by it were in *Business Week*, which took the time to make the musical reference explicit to its terminally unhip readership: "Mash-ups, named after hip-hop mixes of two or more songs, are starting to rock." At the time, a mere sixteen months before the *Business Week* feature on enterprise mashups, the magazine still wasn't sure what to make of the trend. "For now, most mash-ups remain high-tech versions of Tinker Toys," the magazine reported, reaching for another analogy perhaps better suited to its demographic. "After all, how seriously can you take 'Google Map of the Stars,' which zooms in on sites such as Neverland Ranch? But from such whimsical experimentation the next tech blockbuster often emerges."

The blockbuster success of mashups over the following year owed something to Rademacher (whom Google promptly hired) and much to the so-called mashup engines soon offered by Google and Yahoo!. Many companies released code so that their web content could more readily be mashed up by consumers and found that it was a smart business decision: working for free, customers added functionality that, crucially, addressed their actual needs rather than those presumed by marketing teams. Moreover mashed-up sites naturally contributed to one another's web traffic. Even competitors could mutually benefit, as in the case of a Google vs. Yahoo! mashup overlaying Yahoo!'s real-time traffic information on Google Maps.

Because data mashups were contingent on neither advance approval nor financial justification, most of them were decidedly niche applications. For instance, WikiFM linked Last.fm with Wikipedia to automatically display biographies of musicians as their songs were streamed on internet radio. And TypoBuddy applied spellchecking to eBay, alerting users to misspelled auction listings that might draw sparser bidding. Few people were likely to use any one of these applications, but the audience for mashups in aggregate was enormous. Taken together they were a case study in the power of crowdsourcing, collectively making the web a richer environment.

Enterprise mashups were conceived as a corporate corollary to this spontaneous online activity. For example, IBM's mashup portfolio is pitched to companies as "an end to end mashup platform enabling the rapid creation, sharing, and discovery of reusable application building blocks . . . that can be easily assembled into new applications or leveraged within existing applications—delivering lower development costs, rapid ROI, quick delivery of dynamic applications, and greater productivity." In other words, data streams within a company can be combined on the fly, with employees designing their own hybrid software, crowdsourcing their way to higher productivity and bigger profit margins.

If only it were so easy. There has been some limited success to this approach, notably when internal and external sources of data are combined. Several companies, including the enterprise security vendor PGP Corp., have mashed up customer records with Dun & Bradstreet listings in order to improve sales. However, enterprise mashups have yet to become a new corporate paradigm in which employees show the spontaneous initiative of Paul Rademacher, let alone the feral imagination of the

Freelance Hellraiser. As *InformationWeek* put it in 2009, "The fact of the matter is that mashups are really composite applications."

The reason enterprise mashups haven't evolved into anything more radical is that mashing up is an inherently subversive act. Even the most commercially successful mashup deejay, Girltalk, considers the dance floor fodder he produces from mostly mainstream sources to be "a very punk exercise." There may not be anything seditious about his music, but plundering and ravaging the tracks needed to create it requires a spirit of rebellion at odds with corporate organization. *Enterprise mashup* may be an oxymoron.

Still, the very fact that mashups crossed over to the corporate world, complete with off-the-shelf software "enabling the rapid creation, sharing, and discovery of reusable application building blocks," indicates that musical mashups may be reaching their own limits, sounding increasingly formulaic. The most respected mashup artists, such as D.J. Earworm, now mix so many sources, overlaying and juxtaposing them in so many ways, that their work is more accurately described as sound collage. The feedback loop is as nuanced as a conversation, by comparison to which the *A vs. B* setup has the content of a quip.

The fate of the mashup may fall between the artistic and the corporate. As the word drifts toward sound collage on the one hand and composite applications on the other, the most meaningful use of *A vs. B* will most likely be in the mundane creativity of future Paul Rademachers, whose idea of interactivity is hacking the web to make it their own. There will always be a need for the next HousingMaps.com, whether or not the creator bothers to call it a mashup.

K

OK, as abbreviated in txt.

There's an apocryphal story, still in circulation, that the word *OK* was made up by President Andrew Jackson. According to the tale, Jackson used the letters when he was a major general in the War of 1812, marking his approval on papers with initials abbreviating the words *oll korrect*. "The Gen. was never good at spelling," the Boston *Atlas* dryly concluded, recounting the story in August 1840. By that time Old Hickory, as Jackson was known, had served his eight years as president, and his successor, Martin Van Buren, was running for a second term. A native of Kinderhook, New York, Van Buren appealed to the Jacksonian vote with the nickname Old Kinderhook, using the initials O. K. as a political slogan. His Whig Party rivals sought, successfully, to turn his populist appeal into a liability by calling attention to Jackson's alleged semiliteracy. By a sort of logical doggerel endemic in American politics, Old Kinderhook's slogan became a symbol of his ignorance.

The true origin of *OK*, as the American lexicographer Allen Walker Read skillfully uncovered in 1963, was much closer to the *Atlas*'s editorial offices. The letters did stand for *oll korrect*, but the spelling was no accident. The coinage almost certainly came from the waggish editor of the Boston *Morning Post*, Charles Gordon Greene, who was at the center of what Read characterizes as "a remarkable vogue of using abbreviations" beginning in the year 1838. The *Morning Post* was full of them, generally used with a touch of irony, as in the mock dignity of O.F.M. (our first men), or a fit of whimsy, as in the pure zaniness of G.T. (gone to Texas). It was only a matter of months before the fad turned to creative misspelling, a source of humor then as it was in Mark Twain's time.* There was N.C. (nuff said) and N.Y. (no yuse), as well as O.W. (oll wright). The first known appearance of OK followed that pattern. On March 23, 1839, Greene wrote a jocular editorial concerning the Anti-Bell-Ringing Society, naturally referred to as the A.B.R.S., an organization half-teasingly founded to combat a dead-serious ordinance against dinner bells: "He of the Journal, and his *train*-band, would have the 'contribution box,' et ceteras, *o.k.*—all correct—and cause the corks to fly, like *sparks*, upward." This muddle of inside jokes and puns was hardly an auspicious setting for the new word. Nevertheless, as the mania for abbreviation extended to newspapers in New York and beyond, the

---

* One early joke based on misspellings that has endured to this day concerns the three Rs (reading, riting, and 'rithmatic). The jest seems to have originated in England in the early 1820s. By 1828 it had been imported to the United States, where of course it was being attributed to Jackson, as a slur against his education:

    "The three R's—honest 'Rithmetic, Reading & 'Riting"
    I think I can say, I'm no fool in—
    Considering my time was so *took up* in fighting,
    That I only had *three quarters schooling*.

When this ditty started to circulate, Old Hickory was running for his first presidential term. Given the appeal of his backwoods upbringing, the slight probably only bolstered his populist support.

use of *O.K.* also proliferated, until by 1840 it was a readymade plug for Old Kinderhook—and a readymade barb for Van Buren's enemies. The 1840 presidential campaign made *O.K.* a household word and a permanent fixture in the English lexicon, with arguably the widest use in the most languages around the world.*

Recently, though, its lexical hegemony has been increasingly threatened. Spreading exponentially with the multiplication of cell phones, the challenger is, aptly enough, an abbreviation abridging OK itself. "The most common text message must be 'k,'" mused Louis Menand in a 2008 *New Yorker* article. "It means 'I have nothing to say, but God forbid that you should think that I am ignoring your message.'"

As Menand's tone suggests, he isn't a texting enthusiast. He's certainly not alone, and his critique is hardly as obtuse as the philippics of the BBC journalist John Humphrys, who has called texters "vandals who are doing to our language what Genghis Khan did to his neighbours eight hundred years ago . . . pillaging our punctuation; savaging our sentences; raping our vocabulary." Menand has taken the opposite line, condemning texting as "so formulaic that it is practically anonymous," and comparing its richness, unfavorably, to C. K. Ogden's midcentury Basic English.

Perhaps no word supports this viewpoint better than *k.* Though *k* can be used to register nominal assent, most commonly it's a null word, a conversational placeholder perfectly devoid of meaning. Its popularity is largely a function of texting etiquette, most succinctly captured in Norman Silver's "txt commandments." According to the fourth of these, "u shall b

---

* Once it became clear that *OK* was destined to endure, efforts were made to lend it respectability, to disguise its unsavory *oll korrect* origin, by spelling out the acronym phonetically. Even to schoolmarms congenitally allergic to abbreviations, *okay* looked OK.

prepard @ all times 2 tXt & 2 recv." In other words, the act of texting is social, regardless of the linguistic content. Texting in this manner is a return to the grooming practices of our primate ancestors.

And yet the very language of Silver's "txt commandments" suggests a complexity to texting the inverse of Ogden's simplifications. Like so many text messages, his commandments build on creative abbreviations and logograms, combined with inventive capitalization and punctuation, to fit the 160-character limit of the SMS texting protocol. These improvisations, impudent deviations from the language of dictionaries, are presumably what remind traditionalists such as Humphrys of Genghis Khan.* It's a reprise of the confusion about *OK*, another case of cleverness being attributed to ignorance by boors bearing antipopulist sentiments. SMS is not the domain of Genghis Khan, but of Charles Gordon Greene.

The eminent British linguist David Crystal calls texting "ludic" in *Txtng: The Gr8 Db8*, his comprehensive study of texting practices in eleven languages. His careful reading of text messages illustrates this more effectively than any abstract argument: "There are no less than four processes combined in *iowan2bwu* 'I only want to be with you'—full word + an initialism + a shortened word + two logograms + an initialism + a logogram." Teasing out this meaning is a puzzle, and the fun is socially distributed in one person setting up the problem and another solving it.

In the discussion of such word games much has been made of the 160-character restriction on messages, which Crystal compares to the creative challenge of a sonnet, but the sheer number of texts sent—eighty a day for the average teenager—is also an essential factor. Were restrictions on expression all

---

* And Genghis Khan's time was so took up fighting that he didn't *even* have three quarters schooling.

that mattered, the ten-dollar-a-word cost of sending a transatlantic cable in the 1860s would have fostered an intercontinental literary renaissance. Instead nineteenth-century telegraph messages, even those sent shorter distances at more reasonable cost, were reduced whenever possible to a set of numbered stock phrases: *14* for *What's the weather?* and *73* for *Love and kisses.* Lexical and linguistic innovation, in contrast, thrive in settings where modes of communication are so abundant that there's more bandwidth than information. This was true of local newspapers in the 1830s, of which there were an abundance in every city, and it's true of personal electronics today, which effectively allow people always to be connected, and—"u shall b prepard @ all times 2 tXt & 2 recv"—even demands it. By this analysis *k* is not a mark of laxness, but a refrain, such as those reflexively played between solos by jazz musicians.

The apotheosis of idle language is literature, and the literary dimension of SMS has been steadily developing. In 2005 the British cell phone service Dot Mobile began "condensing" classics by canonical authors, from Milton to Shakespeare, to fit SMS format. For instance, Hamlet's soliloquy is neatly rendered as "2b? Nt2b? ???" At around the same time whole novels newly composed on cell phones began to attract a mass audience in Japan. One debut melodrama, by a woman using the pseudonym Mika, attracted 20 million reads on cell phones and computers before becoming a printed bestseller. While many of these novels have the staccato rhythm of communication by SMS, and some use typographic innovations such as emoticons, perhaps the fullest engagement of texting at a lexical level is in poetry. A good example, highly praised by Crystal, is the work of Norman Silver, whose "txt commandments" were in fact composed as a poem and were published in the 2006 collection *Laugh Out Loud :-D.*

However literature is a specialized domain, and what is most remarkable about the language of texting is how artfully it has evolved outside the realm of MFA programs and literary magazines. There is another body of poetry, collectively created and recombined in the eighty-message-a-day traffic between teens, as resonant as folk ballads, albeit in a more contemporary key. Even trite adolescent sentiments take on a typographic flare, as in the following: "f i had a @}—{—4 everytime i thawt of U i wud b wlkN n a @}—{—gRdn 4e." ("If I had a rose for every time I thought of you, I would be walking in a rose garden forever.") However the most poignant line so far to travel by SMS may be one of the simplest: *F?* These two characters can be paraphrased easily enough—"Are we friends again?"—but embedded in the question is the whole dynamic of adolescent relationships in the digital age, where friendship, never simple among teenagers, must be constantly negotiated through multiple channels of (mis)communication. The social need for a phrase as starkly abbreviated as *F?* is matched by the concision with which it has been rendered. *F?* is not at all ludic, and may indeed have developed as an antidote to ludic ambiguity. It is an icon of text's emotional scope and the emotional hold that texting has on those who engage in it.

Louis Menand's *New Yorker* essay, which was composed in response to David Crystal's book, takes issue with Crystal's conclusion that, "for the moment, texting seems here to stay." Menand counters that texting "is a technology nearing its obsolescence," and that the initialisms and pictograms will fade from use as QWERTY keyboards become more common on cell phones. Menand is undoubtedly right that new technologies will change styles of expression. Many initialisms and pictograms will likely go the way of *O.W.* and *O.F.M.*, though some textisms—perhaps *F?*—will migrate to the next platform, just as *OK* outlasted its initial context, and the next one, coming to have nearly a dozen different definitions in the *Oxford English Dictionary*.

What is most notable about texting, though, is the fluidity of lexical invention and reinvention in writing and the fact that such coinage is not in the hands of O.F.M. or Charles Gordon Greene, but rather is truly a popular endeavor. The language of texting, and its legacy, is Jacksonian.

w00t

An exclamation of good fortune.

W00t is a contraction of an exclamation once popular in Dungeons & Dragons, *Wow, Loot!*, imported online by nostalgic gamers. *W00t* is a codeword for *root*, the privileged user account of a system administrator, in the jargon of old-school hackers. *W00t* is onomatopoeic, imitating the sounds made by video games or by Daffy Duck or by railroad trains. *W00t* is an acronym, standing for "We owned the other team" and also "Want one of those." *W00t* originated in dance clubs, where rappers in the early 1990s inspired shouts of "Whoot, there it is!" *W00t* comes from the *Arsenio Hall Show*, the *Wizard of Oz* books, *Pretty Woman, Bloom County*. Or it may be an inversion of the old Scots negation *hoot*, or a perversion of *woeten*, ostensibly an antiquated Dutch greeting.

The many contradictory etymologies of *w00t*, of which the above are only a sampling, have baffled journalists and addled lexicographers since at least 2007, when the dictionary publisher Merriam-Webster announced that *w00t* was the Word of the Year. The term was chosen in a poll conducted on the Merriam-Webster website from a selection of the twenty most popular entries in the company's user-generated Open Dictionary. In a public announcement the publisher was vague about the significance of the vote: "The word you've selected hasn't found its way into a regular Merriam-Webster dictionary yet—but its inclusion in our online Open Dictionary, along with the top honors it's now been awarded—might just improve its chances." And in interviews with the media the Merriam-Webster staff seemed befuddled by the choice. "This is a word that was made up, has no classical roots, but has lasted," the editor at large Peter Sokolowski told *Newsweek*. "I can't say that w00t will stick, but it does show that sense of adventure in language that young people have." In the absence of more authoritative information about the meaning of the word or where it came from reporters cobbled together whatever origin stories they could from the vast repository of lore on Wikipedia, the Urban Dictionary, and the web.

As an antidote to the rumor and myth, the independent lexicographer Grant Barrett published on his Double-Tongued Dictionary blog an essay titled "The Real History and Origin of W00t and w00t," in which he methodically demonstrated that "w00t is, with some caveats, probably derived from and most likely popularized by the dance catch phrase of 1993, 'whoot, there it is!'" He did this by examining the archives of periodicals as well as Usenet and discussion groups, finding a strong correlation between online use of w00t and the rise of two nearly simultaneous 1993 hits, "Whoot There It Is" by 95 South and "Whoomp! (There It Is)" by Tag Team.* Unable to extract any documentary evidence that w00t was used before

---

* As Barrett documented, the producer of "Whoot There It Is" later claimed the term came from the street, and the two members of Tag Team recalled having been inspired by the "Wooof" chant on the *Arsenio Hall Show*.

1993—as D&D lingo, hacker jargon, et cetera—and finding widespread adoption online in that year, Barrett traced this geekiest of cheers to the dance floor.

The response to Barrett's brief essay was fast and furious. The notion that *w00t* originated in pop culture, and dance music in particular, was an insult to gamers and hackers who defined themselves as nonconformist if not anarchic. For them it was self-evident that *w00t* belonged to leet, a semi-encrypted form of English that evolved on Internet relay chat and bulletin board systems in the 1980s to prevent snooping by system administrators. Leet (or l33t) substituted numbers for letters, as in the case of the *oo* in *w00t*, and also modified spellings, as in the case of *w00t* for *root*. Barrett's claim that *w00t* was merely a late mutation of *whoot*, and not the coded celebration of a hacker exploit—*w00t I have root!*—challenged the elitist pretense from which leet took its name.* "It's always amusing to hear an 'expert' tell us what we know from having-been-there-personal-experience is not true," ranted someone with the alias "Regor" in the Double-Tongued comments section, and someone named "Connel" even entered a crazed personal accusation: "I believe Grant fraudulently dismissed the cracker origin of the term just to generate controversy; 100 blog posts = 10,000 page views = increased ad-views for [Barrett's radio show] A Way With Words."

Barrett fought back in terms clearly intended to quell the leetist rebellion. "As someone who was a part of the scene in

---

* *Leet* is short for *eleet*, a leet misspelling of *elite* originally intended to communicate hacker status to other hackers, facilitating the exchange of warez (pirated software) and pron (pornography) without detection by system administrators and their network filters. The leet spelling of *eleet* would actually be *3l33t* or even *31337*. Because the primary purpose of leet was to convey illicit information, and the secondary purpose was to ascend the shadowy hierarchy of a subversive community, leet never had a fixed form. *Hacker* might be written *haxor*, *h4xor*, or even `|-|^><()|z`. One legacy of this unfilterable cipher, familiar to anyone with an email account, is v1aGr4.

the late 1980s and early 1990s, I can assure you that I am well aware of l33t, and I'm hyper-aware of the rubbish theories and provably false origin stories that are spread about it." Yet none of this, nor any of Barrett's impressive lexicographic evidence, addressed a related point brought up by someone called "Decius." "While the connection to 'Whoop, there it is' is interesting, I think focusing on it as the origin of this term is to ignore the forest for a tree," he wrote. "It's like saying that gamers don't say pwn because of the hacker scene, they say pwn because of the ancient origins of the word ownership."

Alongside *w00t*, *pwn* is one of the most popular words in hacker and gaming culture. And like *w00t* it has many origin stories that endure despite the fact that most are as dubious as the connection between *w00t* and Daffy Duck or old Scots dialect. An intentional misspelling of *own*, *pwn* is an emphatic statement of domination, whether of an opponent in a game (*She pwned him!*) or of a computer system (*He has pwnage!*). Folk etymologies range from the historical (*pwn* is an abbreviation of *pawn* coined by MIT hackers developing computer chess in the 1960s) to the graphic (the *p* is an *o* dripping blood) to the exotic (keyboards in some unspecified foreign country lack an *o* key). The most likely derivation, in common with *teh* as a substitute for *the*, is that *pwn* was a common typo, ironically enshrined. Yet that alone does not explain *pwn*'s broad and en-during success as online lingo. The folk etymologies have also played a crucial role, giving a wide range of people, in a vast gamut of circumstances, reason to purposely misspell *own*. For some the connotation is a link to the heritage of hacking; for others the attraction is to a symbol of bloody victory in World of Warcraft or Quest.

What is true of *pwn* is also true of *w00t*. Connections to leet, whether by way of *root* or "We owned the other team," give the word status in certain circles, encouraging allegiance to it that cannot be claimed by *hurrah*, *yippee*, or *yay*. Other folk etymologies, tracing the word back to D&D or even Daffy Duck, resonate for other audiences in equivalent ways.

And that is the underlying reason why *w00t* pwned Merriam-Webster in 2007. *W00t* inspires loyalty like a sports team or political candidate. Not only did people vote for it; they actively campaigned for it. It didn't matter that only a small fraction of the English-speaking population used it or even knew what it meant—far fewer than the number of people familiar with the runner-up, the verb *to facebook*. Partisans of *w00t* were motivated to organize, and they did, edging the word into the mainstream. *W00t's* fallacious folk etymologies provided the roots for lexical legitimacy.

(-///-)

Anime-style emoticon expressing embarrassment.

The snigger point, or note of cachinnation, was invented by Ambrose Bierce in 1887. He proposed the new typographic symbol as "an improvement in punctuation," explaining in an essay that "it is written thus ‿ and represents, as nearly as may be, a smiling mouth. It is to be appended, with the full stop, to every jocular or ironical sentence; or, without the stop, to every jocular or ironical clause of a sentence otherwise serious." Recommended to humorless colleagues who had no trouble recognizing his sarcasm, the snigger point, or note of cachinnation, never caught on.

Similar suggestions have since been advanced, independently, with different motivations. In 1899 the French writer Alcanter de Brahm earnestly proposed that a backward question mark be used in print as a *point d'ironie*, an idea that Alfred Jarry fervently endorsed two years later, though both the irony mark and its creator faded into obscurity shortly thereafter.* And in 1967 *Reader's Digest* ran a short item by the Baltimore *Sunday Sun* correspondent Ralph Reppert, whose Aunt Ev seasoned her family letters with the symbol –) representing "her tongue stuck in her cheek," an idea recirculated on an ARPANET mailing list in 1979 in a proposal to counteract "the loss of meaning in this medium [due to] the lack of tone, gestures, facial expressions, etc." The ARPANET community never embraced Aunt Ev's innovation. But three years later a slightly different icon was rapidly and permanently adopted.

The emblem was first playfully circulated on the Carnegie Mellon University bulletin board system by a computer scientist, Scott Fahlman:

19-Sep-82 11:44 Scott E Fahlman :-)
From: Scott E Fahlman <Fahlman at Cmu-20c>

I propose that the following character sequence for joke markers:

:-)

Read it sideways. Actually, it is probably more economical to mark things that are NOT jokes, given current trends. For this, use

:-(

The :-(symbol was enthusiastically taken up together with the :-), though not in the way Fahlman intended. Instead it was

---

* Even with a greater literary reputation, and the support of figures more mainstream than Jarry, de Brahm's *point d'ironie* probably would not have succeeded, since the introduction of new punctuation is exponentially more difficult than the coinage of new language. The most successful novelty of the twentieth century was the interrobang, superimposing an exclamation point over a question mark to indicate a rhetorical question, which even garnered a coveted key on several models of Remington in the 1960s. The typewriters are now collectors' items.

used to indicate displeasure or sadness, a subtle difference that made all the difference in terms of what happened next: other people began representing more emotions, from affection to frustration to boredom, as sequences of punctuation. Over the following few years, precipitated by the spread of chat rooms and email, joke markers evolved into emoticons.

The number of emoticons now in circulation constitutes a veritable nonverbal lexicon. There are symbols for a kiss :-* and a wink ;-).* An expression of anger :-@ might be met with a plea of innocence o:-) or an expression of surprise :-o or embarrassment :-x. Many emoticons have multiple synonyms. In the case of smileys, eyes may be made with equal signs =-), mouths made with brackets :-], and noses omitted :). Moreover an individual emoticon can sometimes convey more than one meaning. For instance :-/ potentially expresses skepticism, annoyance, discomfort, or indecision. Both of these developments might be expected, given the disparity between the hundred or so standard keyboard symbols and the forty-three facial muscles used in combination to make approximately three thousand emotionally meaningful expressions. Flesh-and-blood smiles and frowns are sometimes misread on account of inattentiveness. In the case of emoticons, beyond the blunt indication that something should be taken humorously :-), there's the added challenge of vagueness :-(. In other words, emoticons often only contribute to the problem they're meant to alleviate: the contextual laxness of slapdash writing fretted about by Alcanter de Brahm :-S and mocked by Ambrose Bierce :-D.

Emoticons have had their greatest success as jokes in their own right, or at least as emblems of playfulness. One form this has taken is caricature, with @:-o portraying Elvis Presley,

---

* The wink may be the salvation of the semicolon, an increasingly marginalized piece of punctuation, just as email saved the @ from typographic oblivion.

(_8^(|) representing Homer Simpson, and ==|:-> standing
for Abraham Lincoln.* A related game involves making animals
and objects, such as fish <'))))) ><< and roses @--->---, the
latter sometimes presented flirtatiously via SMS. And naturally
there are plenty of lewd symbols, such as 8===D to represent
a penis. Easily drawn by anyone, these glyphs are a sort of ASCII
graffiti, an ADD recap of a technique first developed to render
graphics on text-based computers in the 1960s. Given the typically adolescent demographic of ASCII artists, the most popular ASCII images depicted pin-up girls, sometimes using as
many as ten thousand characters to capture nuances of shading,
other times rendering a crude sketch in fewer than two hundred, as in this anonymous tease:

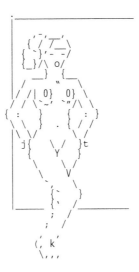

Trading such images through bulletin board systems and Usenet
groups was a way of socializing, albeit in a somewhat antisocial

---

* Like totem poles, these can be stacked to any height, as long as they're only one
character wide. Hence the popularity of emoticons portraying ZZ Top:
B-)=====>

key. Exchanging flowers and faces (and penises) by text message is thus a fitting legacy.

Yet the underlying potential of emoticons has not entirely been lost. Rendered in a different format in Asia, they've achieved a degree of visual expressiveness that has eluded emoticons in the West. "Anime-style" Asian emoticons appear upright rather than tilted to the side, with the face contained in parentheses. As in Japanese anime and manga, much is communicated through the eyes, with (>_<) expressing frustration and (*.*) showing surprise. The configuration also provides space to represent secondary emotional signifiers, such as blushing (-///-) and perspiration (^_^;). Even gestures can be graphically captured, including a wink (^_~), a raised eyebrow (o.O), and tears (;_;). The visual sophistication of these typographic inventions, simultaneously depictive and iconic, is not without precedent. Graphically expressive faces constructed from punctuation were published as early as 1881, when a cartoon in the humor magazine *Puck* printed four of them, representing joy, melancholy, indifference, and astonishment. What anime-style emoticons do is imprint the literary impulse of Alcanter de Brahm with the graphic impact of *Puck*.

Nevertheless Asian emoticons, like their Western counterparts, are most common in text messages and gaming environments, where miscommunication due to carelessness on the part of both sender and recipient is commonplace. Emoticons, even visually compelling ones, are not a panacea for emotional immaturity or negligence. The interesting question is whether, in the right circumstances, for people attentive to their potential, they might become more than mere joke markers or abbreviations for surprise and embarrassment.

One possibility entails the development of an Emotion Markup Language (EML) equivalent to the Hypertext Markup Language (HTML) used as a standard for displaying and linking web pages. As the field of affective computing has evolved to address the mediation of human interaction by machines, EML has drawn academic interest, even leading to the formation of an Emotion Incubator Group by the World Wide Web Consortium. The Emotion Incubator Group has conceived EML as "a 'plug-in' language suitable for use in three different areas: (1) manual annotation of data; (2) automatic recognition of emotion-related states from user behaviour; and (3) generation of emotion-related system behaviour." In other words, the proposed language, which has not yet been implemented on any platform, is intended for internal use by computers. A typical command line might read <category set="everydayEmotions" name="satisfaction"/>. This isn't exactly how people naturally express feelings. However, computer languages tend to become more user-friendly in time. An extended lexicon of emoticons could conceivably encode these emotional states, in which case annotation might become widespread. Emails and text messages might be routinely marked with the mood in which they were written, and this mood might be signaled automatically to the recipient. As technology diminishes face-to-face contact, affective computing might facilitate affective communication.

Ambrose Bierce would surely snigger at the suggestion, considering emotionally marked-up language beneath his literary talent. Were EML limited to joke markers, *points d'ironie*, or notes of cachinnation his opinion would be justified. But emoticons, artfully applied, may eventually open a whole new expressive domain for writing, just as the expressive range of musical notation was expanded several hundred years ago by annotations such as *appassionata* and *furioso* and, the original :-), *giocoso*. More than mere punctuation, the emoticon has the potential to give language a new level of meaning.

Plutoed

Demoted or disqualified.

On August 24, 2006, the ninth planet in our solar system was plutoed by the International Astronomical Union. The scientists did not phrase their decision in those terms. Indeed almost nobody used *pluto* as a verb until January 5, 2007, when the American Dialect Society voted it Word of the Year, prompting worldwide headlines. "Pluto is finally getting some respect," reported the Associated Press, and CNN noted, "Pluto may no longer be a planet, but it has a new claim to fame." It was a good story for a slow news day, which is to say that the brouhaha over *plutoed* was a classic case of truthiness.

Truthiness was the American Dialect Society's Word of the Year for 2005, a term made up by the comedian Stephen Colbert on his satirical TV news program, *The Colbert Report*, to denote something one believes is true based on a desire to believe it even if contradicted by factual information. His primary satirical target was President George W. Bush, but Colbert also meant the word to be a commentary on the media and the

American public. While immeasurably less consequential than Bush's tortuous foreign policy, the anointment of *plutoed* by the American Dialect Society, and the widespread coverage of the selection, can also be explained as a product of wishful thinking. The word was appealing because it was timely.

Yet timeliness alone cannot account for its selection, especially given some of the other timely words with far more legitimate usage, such as *waterboarding*, also nominated. Whereas *waterboarding* was a battle-hardened military euphemism, *plutoed* was mere wordplay, a clever turn of phrase coined for the amusement of language lovers. One of the rare appearances of the term not referencing the American Dialect Society was an article published in the *Montreal Gazette* on January 10, 2007 (a date that nevertheless almost guarantees Word of the Year influence): "The feeling among some was that the ever-maligned Rona Ambrose, who was moved from being environment minister to minister of intergovernmental affairs, really got plutoed in last week's federal cabinet shuffle." Although *plutoed* doesn't add any concrete information about poor Rona Ambrose's demotion, the word charmingly anthropomorphizes the beleaguered dwarf planet in the process of associating human foibles with Pluto's star-crossed plight.

Of course the anthropomorphosis of Pluto predates the Canadian cabinet shuffle. Indeed it explains why the American Dialect Society's Word of the Year selection seemed so apposite to CNN and the Associated Press, and more broadly why the highly technical redefinition of Pluto as a "dwarf planet" was one of the biggest news stories of 2006.

In recent times no heavenly body, except perhaps Mother Earth, has been laden with as much personality as Pluto. Like the other planets in our solar system, Pluto was named after a classical god following its discovery by the astronomer Clyde Tombaugh in 1930, and in keeping with ancient tradition the name was given to capture the planet's character. If Mercury was quick and Venus was pretty, Pluto was cold and dark like the underworld ruled by the deity known as Pluto or Hades.* Yet Pluto's outlier status, reinforced by the fact that Tombaugh was then twenty-four years old and totally unknown, swiftly brought the planet celebrity recognition never accorded to dimly remembered mythological figures such as Uranus and Neptune. The image of Pluto as planetary underdog became permanent in 1931, when Walt Disney introduced Pluto the Pup as Mickey Mouse's hapless pet.

The planet Pluto's fiercely loyal following, especially among American children, was manifest several years before the International Astronomical Union ruling, when New York's Hayden Planetarium eliminated Pluto from a display of planets in 2001. Almost instantly the planetarium's director Neil deGrasse Tyson found himself "a public enemy of Pluto lovers the world over," as he wrote in *The Pluto Files*. The book contains a sampling of the hate mail he received, of which the following, sent by a girl named Madeline Trost, is typical: "Why can't Pluto be a planet? If it's small doesn't mean that it doesn't have to be a planet anymore. Some people like Pluto. If it doesn't exist then they don't have a favorite planet."

Trost's letter succinctly captures the popular sentiment encountered by the International Astronomical Union on August 24, 2006, in response to the release of Resolution 5A:

> (1) A "planet" is a celestial body that (a) is in orbit around the Sun,
> (b) has sufficient mass for its self-gravity to overcome rigid body forces

---

* The name was suggested by an eleven-year-old British girl named Venetia Burney, following global news reports that "Planet X" had been discovered at Lowell Observatory. Planet X was the name given by the observatory founder, Percival Lowell, prior to his death in 1916, based on his hypothesis that Neptune's orbital irregularities were caused by an unknown astronomical body. Tombaugh's discovery elicited dozens of suggested names, including Minerva (suggested by the *New York Times*) and Constance (proposed by Lowell's widow, Constance, who also gamely suggested Percival and Lowell). The most appropriate submission was probably Persephone, since Pluto's eccentric orbit results in the cyclical freezing and thawing of its atmosphere as it transits the sun every 248 years, though this was not known at the time.

so that it assumes a hydrostatic equilibrium (nearly round) shape, and (c) has cleared the neighbourhood around its orbit.

(2) A "dwarf planet" is a celestial body that (a) is in orbit around the Sun, (b) has sufficient mass for its self-gravity to overcome rigid body forces so that it assumes a hydrostatic equilibrium (nearly round) shape, (c) has not cleared the neighbourhood around its orbit, and (d) is not a satellite.

(3) All other objects, except satellites, orbiting the Sun shall be referred to collectively as "Small Solar System Bodies."

The resolution came about as a result of genuine changes to the astronomical landscape. Most notable was the 2005 discovery by an American astronomer, Mike Brown, of a celestial body in the Kuiper Belt that was larger than Pluto. Nicknamed "Xena" (after the cartoon Warrior Princess) the Kuiper Belt Object seemed deserving of being called a planet, as deserving as Pluto, and the IUP's Planet Definition Committee unanimously proposed that it be counted as one. An uproar ensued, as astronomers not on the committee, including Neil deGrasse Tyson and Mike Brown, questioned the wisdom of conferring planetary status on Kuiper Belt Objects. Soon there would be hundreds of planets, they predicted, as improved optics detected increasingly many orbiting spheres of rock and ice. To prevent that from happening, the IAU General Assembly rather arbitrarily insisted that a planet not only be round and in orbit around the sun, but also that it be substantial enough to have cleared a path for itself. Xena hadn't done that. Nor, as it happened, had Pluto. Both were dumped into the dwarf planet ghetto.[*]

---

[*] Originally members of this new category were given the less derisive name *pluton*, but the word was discovered to have another meaning in geology. After several other names were rejected, including *planetino* and *Tombaugh object*, the International Astronomical Union finally settled on *plutoid* in 2008, a term defended (rather unconvincingly) by IAU president Catherine Cesarsky as a boost to Pluto's standing, making it a "prototype of a new type of fascinating objects."

The exclusion of Pluto from the planetary pantheon not only offended schoolchildren; it also rankled some politicians. The New Mexico State Legislature passed a resolution in 2007 stating that, "as Pluto passes overhead through New Mexico's excellent night skies, it be declared a planet." However, the most substantive criticism came from within the ranks of astronomy, where Alan Stern, leader of NASA's New Horizons mission to Pluto, fomented a veritable rebellion. "This definition stinks, for technical reasons," he ranted to *Space.com* on the day that the IAU resolution was passed. Those technical reasons included the fact that four of the remaining eight planets—Earth, Mars, Jupiter, and Neptune—have not cleared their orbits of asteroids and should therefore also logically be excluded from planethood.

Yet the underlying problem was ignored by both factions: whether inclusive of Pluto or not, all current definitions are premised on an ancient Greek notion of *planetes asteres*, or wandering stars, which included the sun and moon in addition to modern planets such as Mars and Jupiter. In other words, the original definition was perceptual rather than ontological. It concerned how we saw the night sky, not what was actually there. Planetary status was a question not so much of truth as of truthiness.

And truthiness is what is fundamentally at stake today, on both sides of the planetary divide. Many people believe Pluto ought to be considered a planet, all the more so given its underdog appeal. For many others the idea that there might be hundreds of planets rather than merely nine (or eight) is an affront to planets' special place in Western culture. Both inclinations are sentimental, even if backed up by scientific data, because the whole idea of wandering stars is a matter of beliefs, not facts. The most sensible solution would be to pluto the word *planet* from the scientific vocabulary and establish a new lexicon based on modern scientific understanding. One term might be assigned to any object with sufficient gravity to round itself out, and another might designate whether it travels alone in its orbit. *Planet* would be left to stargazers whose relationship to the night sky is akin to the ancient Greeks'.

That will not happen. Tradition counts for more than reason in matters of language, and as the brief life of *plutoed* suggests, vocabularies are rarely shaped by the will power of lexicographers. (Even the use of *truthiness* nearly vanished with the Bush administration.)

Mike Brown captured the true nature of the Pluto debate better than anyone when he finally gave Xena a permanent name following the IAU resolution. The name he chose was Eris, after the Greek deity who instigated the Trojan War by tossing a golden apple to her fellow goddesses and declaring that it belonged to the fairest. "Eris caused strife and discord by causing quarrels among people," Brown explained to Reuters, "and that's what this one has done too." Coveted as senselessly as a golden apple, planetary status is as subjective as the fairness of goddesses.

PART VI

NEOLOGISM

The most popular sedative in the early twentieth century was the chemical potassium bromide. Around 1906 the American humorist Gelett Burgess observed that there were people who could put him to sleep at least as effectively as the drug. He dubbed them *bromides* and referred to their tiresome clichés—"Every dog has his day"—as *bromidioms*.

Burgess also proposed a lot of other neologisms, including *floogijab*, an insult masquerading as a compliment, and *oofle*, someone whose name you've forgotten. These might have made people chuckle (as did his notorious poem, "The Purple Cow"), yet they were too self-conscious to enter the English language. The same was true of *bromidiom*. Only *bromide* survived, broadening also to encompass the clichés themselves and enduring long after the medical industry discovered new sedatives.

From this brief history some generalizations can be made, which may help foretell the fate of newer neologisms, such as *singularity* and *spime*. Burgess's *bromide* could quite plausibly have come into the language unassisted. The etymology provides a pedigree (or at least an alibi), and the sound and look of *bromide* are not noticeably contrived.

But in the unruly domain of human expression every dog has his day and every generalization calls for exceptions. Burgess is also famous for another widely used neologism, which derives from a make-believe character he drew on the cover of a book he wrote about bromides. Her ludicrously buxom figure was an obvious satire of the tacky packaging publishers used to flog books. He gave her the most ridiculous name he could think of. He called her Miss Belinda Blurb.

SINGULARITY

The technological infinite.

In the future humans will live forever. Bodies will be optional. Brains will be networked, and the whole universe will be sentient. All of this is an inevitable consequence of the singularity, the moment at which computers surpass human intelligence. And the singularity will inevitably occur by the year 2045.

These are the predictions of Ray Kurzweil, one of the most successful and respected technologists alive, the father of speech- and optical-character-recognition software, fundamental advances in artificial intelligence that have tapped computers into the two primary modes of human communication: oral and written. Both breakthroughs, conduits of machine learning, were achieved by imbuing computers with basic pattern recognition, such that voices with different accents and alphabets in different typefaces could be deciphered. Key to our knowledge and understanding, pattern recognition is something that humans are very good at, and Ray Kurzweil is better than most people at recognizing patterns. Perceiving subtle connections between varied technologies, he has foreseen the victory of a computer over the world's leading chess player as well as the proliferation of seminal technologies, including the world wide web. These are reasons that prominent figures from Bill Gates to Marvin Minsky take Kurzweil seriously.

Kurzweil's prediction of the singularity follows his pattern of pattern recognition, enlarged to the scale of all history. He claims that he first recognized it while plotting human advances from the wheel to the web, discovering a curve of technological advancement that was not linear but exponential. He dubbed this the Law of Accelerating Returns and hired a team of researchers to trace it back to the Big Bang. Then he began ploddingly to move forward in time, until he reached "a profound and disruptive transformation in human capability," circa 2045. At that point, he writes in *The Singularity Is Near*, his best-selling 2005 book on the subject, "the entire universe will become saturated with our intelligence. This is the destiny of the universe."

Kurzweil was not the first to make this observation. But "it wasn't very quantified" before he began charting the past 13.7 billion years of cosmic progress in the mid-1990s, as he told *Rolling Stone* in a 2009 interview. "It was more impressionistic."

Actually in the beginning it was strictly metaphoric. The first recorded use of the word *singularity*, in terms of technology, was by the mathematician John von Neumann, as his colleague Stanislaw Ulam recalled in a tribute published a year after von Neumann's death in 1957. "One conversation," wrote Ulam in the *Bulletin of the American Mathematical Society*, "centered on the ever-accelerating progress of technology and changes in the mode of human life, which gives the appearance of approaching some essential singularity in the history of the race beyond which human affairs, as we know them, could not continue." A quarter-century later the science fiction writer Vernor Vinge elaborated on the image—readily acknowledging his debt to von Neumann—applying the word to an idea proposed by the statistician I. J. Good in the 1960s: that an ultraintelligent machine would be able to build a machine even more intelligent, ad infinitum, resulting in an "intelligence explosion." In a 1983 column for *Omni* magazine discussing the challenge of writing sci-fi projected even into the near future, Vinge predicted that people "will soon create intelligences greater than our own. When that happens, human history will have reached a kind of singularity, an intellectual transition as impenetrable as the knotted space-time at the center of a black hole, and the world will pass far beyond our understanding."*

Vinge's description departs from von Neumann's in a few essential respects. The first, as he himself pointed out when he revisited the idea a decade later, was that von Neumann appeared to be "thinking of normal progress, not the creation of a superhuman intellect" in the sense that Good first advanced. The second, perhaps subtler, is that von Neumann's metaphor seems to be purely mathematical (i.e., a point at which the value calculated for a function is infinite), whereas Vinge's is overtly cosmological (i.e., a point of infinite density at the center of a black hole). A specific case of mathematical singularity arising from the equations of general relativity, a cosmological singularity has the advantage for a science fiction writer of being visceral, and for a prognosticator of being unknowable.

Vinge's idea remained on the fringe, lapping at the edges of his fiction, until 1993, when he was invited to present a paper, "The Coming Technological Singularity," at the VISION-21 Symposium, sponsored by the NASA Lewis Research Center and the Ohio Aerospace Institute. He made his position clear in the very first sentences: "The acceleration of technological

---

* Vinge claims to have first used the term with reference to superhuman intelligence a year earlier, at the annual conference of the Association for the Advancement of Artificial Intelligence, but no transcript records what he said there.

progress has been the central feature of this century," he argued. "We are on the edge of change comparable to the rise of human life on Earth. The precise cause of this change is the imminent creation by technology of entities with greater-than-human intelligence." The boldness of Vinge's claim was backed by lucid description of the mechanisms that might lead to such a scenario, including consciousness arising in artificially intelligent machines and computer enhancement of the human brain. He set the transformation at thirty years. "Shortly after," he said, "the human era will be ended."

The singularity metaphor remained central to his vision yet also took on new implications. "It's fair to call this event a singularity," he said. "It is a point where our old models must be discarded and a new reality rules, a point that will loom vaster and vaster over human affairs until the notion becomes a commonplace. Yet when it finally happens, it may still be a great surprise and a greater unknown." What is extraordinary about Vinge's paper is his ambivalence—poised tenuously between excitement and trepidation. Humbling in its unknowability, Vinge's singularity is cause for self-reflection in the present.

"The Coming Technological Singularity" has often been credited with launching a so-called Singularitarian* movement. Widely distributed, Vinge's creed was certainly more broadly read than Ulam's reminiscences in the *Bulletin of the American Mathematical Society*. Yet the appeal of the paper was primarily philosophical. Vinge's singularity was less something to rally

---

* Ray Kurzweil credits the term *Singularitarian* to the Extropian Mark Plus, who in 1991 defined it as "one who believes the concept of a Singularity." (The Extropians were a sort of proto-Singularitarian movement that sought "to overcome human limits," achieving eternal life and infinite intelligence through technology. In this context *Mark Plus* was the pseudonym of a member named Mark Potts.) In *The Singularity Is Near* Kurzweil assumes the Singularitarian mantle and expands on Plus's definition, describing a *Singularitarian* as "someone who understands the transformations that are coming in this century and who has reflected on the implications for his or her own life." His book scarcely even mentions Vinge.

behind than to ruminate about.* To become truly popular the singularity needed a champion. And that, more than anything else, is the role taken by Ray Kurzweil. As a 2008 profile in *Wired* magazine phrased it, "Kurzweil transformed the singularity from an interesting speculation into a social movement."

Kurzweil's version of the singularity is dazzlingly elaborate and unequivocally optimistic. He supports his Law of Accelerating Returns not only with vast graphs of multicentury trends but also with catchy examples, such as the Human Genome Project, which haltingly analyzed only one ten-thousandth of the genome in the first year, yet successfully delivered the entire sequence a decade and a half later. And in interviews with journalists, he shows off his cell phone, a device "a million times cheaper and a thousand times more powerful than the computer that [they] used at MIT when [he] was an undergraduate," as he told *Newsweek* in 2009. "That's a billionfold increase. And we'll do that again in the next 25 years." Although this is just a tangible example of Moore's Law,† his

---

* Even for those dismissive of Vinge's conclusions, his singularity affords occasion for deep intellectual mediation and retains value as a thought experiment, in the spirit of the finest science fiction. One reflection on life after the singularity: "Immortality (or at least a lifetime as long as we can make the universe survive) would be achievable. But in this brightest and kindest world, the philosophical problems themselves become intimidating. A mind that stays at the same capacity cannot live forever; after a few thousand years it would look more like a repeating tape loop than a person. . . . To live indefinitely long, the mind itself must grow . . . and when it becomes great enough, and looks back . . . what fellow-feeling can it have with the soul that it was originally?"

† Initially proposed by Intel founder Gordon Moore in 1965, Moore's Law predicts that the number of transistors on a computer chip will double every two years. This prediction has held true for an impressive three and a half decades, but most technologists, including Moore, are skeptical that it will apply in the future, as components become so small that they're upset by quantum effects. Moore is also strongly doubtful that there will ever be a singularity, a skepticism shared by many other technologists whose innovations are often invoked by Singularitarians. For instance, Jeff Hawkins, a pioneer of handheld computing, has insightfully observed in the industry journal *IEEE Spectrum*, "Exponential growth requires the exponential consumption of resources (matter, energy, and time), and there are always limits to this. Why should we think intelligent machines would be different?"

rhetorical spin casts it as a universal truth in which we're somehow, inevitability aside, active participants.

Based on this heroic past Kurzweil promises a future in which nanotechnology will convert a simple rock, according to *The Singularity Is Near*, into "a cool, zero-energy-consuming computer with a memory of about a thousand trillion trillion bits and a processing capacity of 10^{42} operations a second, which is about 10 trillion times more powerful than all human brains on Earth." He calls this transformed substance *computronium* and imagines such computational power spreading, perhaps faster than the speed of light, to convert all the matter and energy of the universe into an omniscient artificial intelligence. "We will determine our own fate rather than have it determined by the current 'dumb,' simple machinelike forces that rule celestial mechanics," he vows. The cosmos will become conscious.

Kurzweil's singularity has the scope of a religion—"the rapture for geeks" in one popular formulation—and shares with most cults ample opportunity for personal participation. There are dietary supplements to be taken in order to live through 2045, on the other side of which lies the promise of immortality. (Kurzweil ingests between 180 and 210 pills a day. He also sells them through Ray and Terry's Longevity Products, where a one-month supply of "Antiaging Multipack" sells for $86.75.) There are also preparatory courses in robotics and bioinformatics and nanotechnology at Singularity University, cofounded by Kurzweil at the NASA Ames Research Center. (Tuition for a nine-day "Executive Program" is $15,000.) In addition to generating income these quaintly low-tech offerings—capsules and classes—help to make the impending singularity routine, a calendar item instead of a philosophical conundrum.

Through this transition from von Neumann's figure of speech and Vinge's description to Kurzweil's product line, little more than the name remains the same. That it persists is testament to the singularity's evocative power. As every schoolchild knows, even light cannot escape a black hole. The singularity at its center is inherently shrouded in mystery. A singularity is a point at which science meets mysticism.

Yet a metaphor, even a vivid one, can withstand only so much elucidation, and Kurzweil's urge to comprehensively describe the universe after 2045 strips away much of what makes the black hole comparison apt. He insists on the cosmic imagery. "Just as a black hole in space dramatically alters the patterns of matter and energy accelerating toward its event horizon," he writes on the opening page of *The Singularity Is Near*, "this impending Singularity in our future is increasingly transforming every institution and aspect of human life." But celestial objects less exotic (stars, for instance, even planets) dramatically alter the patterns of matter and energy; moreover the effect of a black hole is merely local. As Singularitarian ambitions have overwhelmed passive pattern recognition and come to encompass nothing less than universal dominion, the singularity metaphor has atrophied into a fanciful phrase as callow as a political slogan. And the vulgarizing of the word signals the vulgarizing of the conception—fantastically impressionistic— once associated with it.

On the other hand metaphors and the impressions made by them may hide vagueness or confusion, and Kurzweil's singularity, stripped naked, does call attention to the core vulnerability of Singularitarian thinking. In mathematics singularities are sometimes called "pathological," since an infinite value for a function often indicates an error. This is especially the case where functions model real-world scenarios, such as the description of space-time within a black hole. For this reason black holes have come to be seen as a physical challenge to the equations of general relativity, an indication that relativity must be modified or folded into a new theory, such as loop quantum gravity. Likewise the infinite promise of the technological singularity starts to seem pathological once the event horizon of 2045 is obliterated, rocks become more intelligent than the world population, and even the intractable speed of light gives way to the dawning consciousness of the cosmos.

As the mathematician and physicist Henri Poincaré observed in 1899, "Logic sometimes makes monsters." If ever there are ultraintelligent machines, these monsters will be their nightmares—just as our nightmares are metaphors.

Quid

Quasi-universal intergalactic denomination.

Of the many challenges facing tourism in space, one of the least obvious is the problem of intergalactic monetary exchange. Far more pressing to the nascent industry are issues such as extra-terrestrial transportation and gravity-free accommodations. Charles Simonyi's twelve-day trip to the International Space Station in 2007 cost him $25 million, more than the budget of an average family vacation. Yet years before even the most optimistic technophiles expect space tourism to be more than a fifteen-minute suborbital joyride on Virgin Galactic, a currency has been established, initially trading on Travelex for $12.50. It's called the *quid*.

Quid is an acronym for "quasi-universal intergalactic denomination." Of course it's also an appropriation of British slang for the pound sterling, and it is this association with the common term for a familiar item that gives it resonance, an evocative word for a provocative concept.

One might have expected the new space money to repurpose the official name of an existing currency. The British and French have preferred that strategy when they've colonized other countries, and even Douglas Adams, for all his creativity, fell upon the formula when he coined the *Altairian dollar* in *The Hitchhiker's Guide to the Galaxy*. But colonization robs a place of its exoticism. And if space tourism has any purpose, it's escapism in extremis.

Unlike the pound or the dollar, the quid has no inherent allegiances. The word has also been used at various stages as slang for the shilling, the sovereign, and the guinea, as well as the euro and the old Irish punt. Even the origin is "obscure," according to the *Oxford English Dictionary*, which cites a characteristic early use of the word in Thomas Shadwell's *Squire of Alsatia*: "Let me equip thee with a Quid." The 1688 publication date of Shadwell's play overrules one popular folk etymology, which claims that *quid* is short for Quidhampton, location of a mill that produced paper money for the Bank of England. The Bank of England wasn't established until 1694. Most other historical explanations also seem dubious: *quid* is an old term for a mouthful of tobacco, for instance, a variation of *cud* dating back to the eighteenth century, but a sovereign or even a shilling seems a steep price to pay for a cheek of nicotine. That leaves Latin, in which *quid* means "what" or "why" and is most familiar in the legal expression *quid pro quo*, designating an exchange in the abstract, literally translated as "what for which."

Such abstraction is appropriate given the unknowns of shopping in space, and Travelex offers no notion of what this novel quasi-currency might buy, focusing instead on the certainties of interplanetary travel, all scrupulously applied to the fabrication of quid. The design is based on research undertaken by scientists at Britain's National Space Centre and the University of Leicester, who selected the polymer polytetrafluoroethylene, otherwise known as Teflon, for its nonreactive chemistry and shaped the tokens as lozenges to prevent them from nicking fragile equipment in zero gravity. The scientists also considered electronic currencies, but chips were deemed vulnerable to cosmic radiation, and credit cards connected to terrestrial banking systems were judged unreliable at light-year distances. Minted on Earth for more than two and a half millennia, coins seem to be an optimal technology in space.

Yet at this point the specie counts for less than the idea, and it is the circulation of the word *quid* that has real value. Given a term for fungible space currency people can talk about more than seductive technologies and begin to consider the social and cultural implications of extraterrestrial living. Granted, deep space tourism remains in the realm of science fiction. Nevertheless, in contrast to *The Hitchhiker's Guide to the Galaxy*, sci-fi becomes participatory when the quid is discussed and debated. Language becomes a probe of our future, rocketing us toward remote possibilities eons before they become inevitabilities or impossibilities.

It's appropriate that *quid* means "what," and even more so that it means "why." Marginally a financial instrument, the quasi-universal intergalactic denomination is essentially a philosophical token.

Spime

A hypothetical self-aware object.

The word *robot* first appeared in print in 1920, forty-one years before robotics became an industrial reality. Derived from the Czech term *robota*, meaning "forced labor," the name was given to the automata in *R.U.R.*, a play by Karel Čapek in which machines manufactured by humans eradicate their creators. When the play traveled to the United States in 1922, the *New York Times* called it "a Czecho-Slovak Frankenstein." Isaac Asimov was somewhat less charitable in a 1979 essay: "Capek's play is, in my own opinion, a terribly bad one, but it is immortal for that one word. It contributed the word 'robot' not only to English but, through English, to all the languages in which science fiction is now written."

Asimov was himself one of the foremost authors of this genre, coining the word *robotics* in a 1941 story, and nine years later formulating the Three Laws of Robotics,[*] the first code of

---

[*] The Three Laws of Robotics are as follows: "1—A robot may not injure a human being, or, through inaction, allow a human being to come to harm. 2—A robot must obey the orders given it by human beings except where such orders would conflict with the First Law. 3—A robot must protect its own existence as long as such protection does not conflict with the First or Second Law."

conduct for machines. Those laws have immeasurably influenced real-world engineers in the decades since the first working robot, the four-thousand-pound Unimate, was installed in a General Motors plant in 1961, just one example of how Čapek's immortal word, freed of its trite theatrical frame, has profoundly impacted the evolution of technology.

The fate of *R.U.R.* stimulates a provocative question: Can an effective work of science fiction be written in a single word?* At least one seems worthy of consideration. That word is *spime*.

Spime was coined by Bruce Sterling, a Hugo Award–winning author of numerous sci-fi novels that have helped to define cyberpunk. Many of those novels, such as *Heavy Weather* and *Holy Fire*, are set in the near future, presenting dystopic visions of what our world might become if we continue to behave as irresponsibly as we have in the past. *Heavy Weather*, for instance, is a story of a globally warmed environment ravaged by tornadoes, one of which threatens to devastate the planet unless "hacked" by a cyberpunk Storm Troupe. And *Holy Fire* describes a "gerontocracy" populated by technologically enhanced humans able to survive indefinitely, though at the expense of their humanity. For all their originality these books are conventional in their novelistic exposition of ideas. Sterling's *heavy weather* and *gerontocracy* belong to the same narrative tradition as *robot* and *robotics*.

Though equally fictitious, spimes have no such fictional scaffold. The first use of the word was on August 9, 2004, at the SIGGRAPH computer graphics convention in Los Angeles, where Sterling was the keynote speaker. He began his lecture by talking about *blobjects*, a name given by the industrial designer

---

* Stanisław Lem may have been closing in on this reductionist ideal with *A Perfect Vacuum*, his 1971 collection of metafictional essays reviewing science fiction novels that had never been written.

Steven Skov Holt to objects with bulbous forms, from sofas to automobiles, visibly conceived on a computer. Blobjects had a history reaching back to the late 1980s, when Holt coined the term in *ID* magazine, yet for Sterling they seemed merely a primordial stage in the development of a whole new ontological category. Blobjects imbued physical things with the sleek look of computer graphics. Sterling envisioned objects that would also *behave* like animations. "Having conquered the world made of bits, you need to reform the world made of atoms," he exhorted the legions of designers in the SIGGRAPH audience, many preeminent in their field and all seeking the next big thing. "Not the simulated image on the screen, but corporeal, physical reality. Not meshes and splines, but big hefty skull-crackingly solid things that you can pick up and throw. That's the world that needs conquering. Because that world can't manage on its own. It is not sustainable, it has no future, and it needs one."

Sterling proposed that this could be achieved by a natural extension of existing technologies such as global positioning systems, radio frequency identification, and WiFi networks. He predicted that these technologies, integrated with appropriate sensors and software, could make a sort of product that was, in a manner of speaking, self-aware. "We can call it a 'Spime,' which is a neologism for an imaginary object that is still speculative," he explained in his speech.[*] "The most important thing to know about Spimes is that they are precisely located in space and time. . . . Spimes have identities, they are protagonists

---

[*] While Sterling mostly capitalized the word *spime* in the text of his speech, which was subsequently posted on *BoingBoing*, his first published article about spimes, in *Wired* two months later, set the word in lowercase. This was a crucial step in the word's evolution, signifying a renunciation of proprietary ownership. Thereafter the meaning of *spime* would be open-source, coauthored by all who used it, like any word in the dictionary. The word *robot* underwent a similar transformation, having been capitalized in Čapek's original text, becoming lowercase in general use.

of a documented process. They are searchable, like Google. You can think of Spimes as being auto-Googling objects."

If the concept was simple, the implications Sterling drew from it were deep. Because spimes are networked, and information is shared, he surmised that "a spime is a users group first, and a physical object second." And because spimes record all that has happened to them from the manufacturing process forward, "a Spime is today's entire industrial process, made explicit. . . . A Spime is an object that ate and internalized the previous industrial order." Acknowledging the complexity of his description, Sterling argued that spimes were nevertheless essential because only objects with this sort of self-awareness would be able to guide themselves through the cycle of consumption and disposal. He believed that spimes were needed, as he wrote in *Wired* magazine, "to sustain civilized life in the long run."

Half a decade later the world made of atoms still had not been reformed, as Sterling acknowledged in his blog, *Beyond the Beyond*, on March 26, 2008. Responding to a programmer's non sequitur use of *spime* to describe a virtual bacterium he'd generated in Second Life, Sterling noted, "Spimes are speculative, they do not yet exist," and he argued that "the idea and description of 'spimes' will likely sound hopelessly outdated well before spimes become practical." Yet the Second Life coder was hardly alone in setting the word to use. In fact several years before, Sterling had felt compelled to start a "Spime Watch" category on *Beyond the Beyond* and by 2008 had collected hundreds of examples. Many of them, such as a pair of GPS-outfitted shoes for Alzheimer's patients and a tag to track the afterlife of trash, didn't explicitly reference the word *spime*, but others did, such as the OpenSpime internet protocol and a "location intelligence" company called Spime, Inc. These approximations, muddled though they may be, have only made Sterling's vision more palpable. And the word has provided a way of talking about what all of them might mean, integrated, in the future.

The speculative genesis of *spime* at SIGGRAPH, and the ongoing observation of the concept on Spime Watch, have given the word a quality missing in *gerontocracy*, which remains rooted in *Holy Fire*, and a momentum lacking in Sterling's other naked neologisms, such as *buckyjunk*, coined in a 2005 *Wired* essay describing the sludge of buckyball nanowaste he believes will be the byproduct of nanotechnology. Without question this is partly on account of the term's imaginative merit: *spime*, like *robot*, is considerably more visionary than *gerontocracy* or *buckyjunk*. The multiple channels through which the word has propagated have also played an important role. (*Robot* likewise benefited from exposure in multiple media, including BBC television and radio broadcasts of the play, in addition to the book and stage production, before migrating into other authors' fiction.) Spimes occupy a twilight zone between fictional narrative and physical object. They are a collective hypothetical.

Hypotheticals are typically the business of futurists, who often coin words to make them memorable. In his influential books Alvin Toffler introduced terms such as *cognitarians*, the proletarians of the impending "Third Wave" information economy. Far more frivolous, Faith Popcorn annually bombards newspapers with catchy predictions about forthcoming trends, such as *manity* (male vanity), which bring business to her marketing consultancy. Yet these terms, presented as the product of scientific prediction, do not belong to the creative domain of fiction. They bear some resemblance to Sterling's *buckyjunk*, but they aren't in the same sublimely speculative category as *spime*.

If spimes have an ancestry, it is in the "anticipatory design" of R. Buckminster Fuller, his Dymaxion House in particular. The first instance of prefabricated housing, the Dymaxion House didn't exist when newspapers started publishing articles about it in 1929. Moreover many of the requisite plastics and alloys and photovoltaic cells hadn't been invented, and his system of delivery by dirigible was entirely fanciful. At the time there was only Fuller's description, delivered at great length to anyone who would listen, and a scale model installed at the Marshall Field's department store in Chicago. Nevertheless Fuller presented his house with the sober assurance that it would be built, and even though it never went into production his model served as a fantastic premonition of modern construction. (As *Time* magazine wryly noted in a 1964 cover story about Fuller, his design "carried Corbusier's 'machine-for-living' concept farther than the Continental avant-garde had dared to think it.") Though Fuller would never have admitted it, the unbuildable Dymaxion House was a work of science fiction in architectural form, as the spime is in lexical form.

A harbinger of technology to come, even if that technology is unrecognizably different from Bruce Sterling's vision, *spime* is a work of anticipatory language.

Exopolitics

Foreign affairs with alien races.

At the Dwight D. Eisenhower Presidential Library, an archivist named Herb Pankratz specializes in queries about the thirty-fourth president's exopolitics. Pankratz assumed this responsibility because of his expertise in transportation. Since exopolitics involves diplomacy with visitors from other planets, his colleagues deemed him the best qualified person on staff to field questions, of which there are many, since Ike is alleged to be the first president to have negotiated directly with aliens.

Neither Pankratz nor anyone else at the Dwight D. Eisenhower Presidential Library is able to confirm these historic events. They tell researchers that the president's presumed first meeting with extraterrestrials, on the evening of February 20, 1954, was in fact a dental appointment. They inform people that Eisenhower's emergency departure from the Smoking Tree Ranch, where he was vacationing, was on account of a chipped porcelain cap on his upper right incisor, broken when he bit down on a chicken bone, not a secret meeting at Edwards

Air Force Base with aliens requesting that he end America's nuclear weapons program in order to protect the space-time continuum. The archivists have no record of the words with which Ike rebuffed his celestial guests without causing an intergalactic diplomatic rift, nor of the accord he allegedly reached with a different alien race later that year, allowing them to borrow cows and humans for purposes of medical examination, provided that they return the specimens unharmed.

The lack of documentation has not been taken as want of evidence by organizations such as the Exopolitics Institute. On the contrary the information gap has only further convinced them of a governmental cover-up, extending to the present day, a "truth embargo" involving not only Secretary of State Hillary Clinton but also the United Nations.

As a conspiracy theory exopolitics is barely worthy of a B movie. Even without asking how these extraterrestrial ambassadors have avoided public exposure for over half a century, one might legitimately wonder why the world's governments have so persistently hidden them, and why beings of allegedly superior intelligence have proven so complacent. The most popular explanation for this "cosmic Watergate" is that the aliens possess valuable technology, generally involving energy, which governments want to hoard for purposes of world domination. If so, one might legitimately ask, more than fifty years after Ike chipped his tooth on a chicken bone, what exactly his successors are waiting for.

Yet exopolitics has thrived as a nexus for conspiracy theories since the term was popularized in an e-book written by the "space activist" Alfred Webre in the year 2000. The Exopolitics Institute, a "501(c)(3) non-profit educational organization dedicated to studying the key actors, institutions and political processes associated with extraterrestrial life," publishes the online

Exopolitics Journal, featuring articles such as "Deep Politics, End-Games and Agendas: How Exopolitics Can Offer Avenues to Resolve Population Reduction and Other Eco-Conundrums." The Exopolitics Institute also provides instruction for would-be exopoliticians. For $1,200 tuition one can learn "the conceptual skills and diplomatic training" needed to practice "citizen diplomacy in extraterrestrial affairs," earning a Galactic Diplomacy Certificate. A few real politicians, such as former Canadian defense minister Paul Hellyer, have even been enlisted to lend credibility to exopolitical conventions, catching the attention of mainstream periodicals, including the *Washington Post*. But the best explanation for the persistence of exopolitics may be the name itself.

The term *exopolitics*, an extension of *geopolitics*, is built on the same linguistic pattern as *exoplanet* and *exobiology*. Even casual readers of science will be familiar with the former since planets orbiting distant stars have been detected by space telescope, making headline news. Some of these exoplanets are rocky like Earth, giving momentum to exobiology, the study of nonterrestrial life or, more precisely, what forms such life might hypothetically take. At least superficially, particularly in our search engine culture, exopolitics looks like just another of these exotic areas of research. For instance, Yahoo! Search Assist suggests *exopolitics* just a few items below *exoplanets* when one enters their shared initial letters.

Of course no scientist will be convinced by the linguistic camouflage, and most laymen will also be skeptical after brief perusal of the supposed truth embargo on diplomatic relations with aliens. The trouble for science is that the association also runs in the other direction: just as the perfectly plausible idea that there might be life elsewhere in the universe has been made ridiculous by the popular image of little green men, exopolitics makes exobiology seem precariously fringy.

In fact exobiology is about as fringy (or exotic) as a dental appointment. Much of it involves the examination of extremophile bacteria found in places such as the Arctic and Death Valley or in nuclear waste dumps in order to map out the full gamut of life on Earth and the environments that can sustain it. Such information may help to narrow down where we might productively look for extraterrestrial life, and equally important, how we might detect it with a space probe or inside a meteorite.

However, there is another side to exobiology, more philosophical. Recognizing extraterrestrial microbes may depend on a broader notion of life than comes naturally to us, even as we extend our realm of study to the farthest reaches of Earth. Independently evolved, extraterrestrial life might not be based on DNA or amino acids or even carbon. Embedded in mundane diagnostics for testing Martian rock is a profound question: *What is life?*

For all its fringiness exopolitics is equally, similarly provocative. We don't know whether there are intelligent beings elsewhere or whether we'll ever communicate with them, let alone need to establish intergalactic diplomatic ties. But simply to inquire—without the conspiracy theories—what relationship we might have with independently evolved minds is a basis for asking a question as profound as the foundational question of exobiology: *What is humanity?*

Panglish

A simplified future world English.

"I am of this opinion that our own tung should be written cleane and pure, unmixt and unmangeled with borowing of other tunges," wrote Sir John Cheke in 1561, defending English against the deluge of language imported from French and Italian. The first professor of Greek at Cambridge University, Cheke did not object to foreign phrasing out of ignorance, but rather argued from principles so fastidious that his translation of the Gospel According to Matthew substituted the word *crossed* for *crucified* and *gainrising* for *resurrection*. Proud of his heritage, unbowed by European cultivation, Cheke refused to be indebted to other cultures in his expression, "wherein if we take not heed by tiim, ever borowing and never paying," he warned, "[our tung] shall be fain to keep her house as bankrupt."

Nearly half a millennium has passed, and Cheke's disquiet seems ridiculous, not only because English has been incalculably enriched by mortgaged non-Germanic words such as *democracy* and *education* and *science*, but also because our own tongue has so flourished as to be seen on the European continent and around the world as the sort of cultural threat that Classical and Romance languages were to Cheke's countrymen. The predominance of English is staggering. An estimated 1.5 billion people speak it, a number that the British Council predicts will increase by half a billion by the year 2016. Moreover fewer than a quarter of these people speak English as a first language; there are nearly twice as many nonnative speakers in India and China as native speakers on the planet.

As might be expected given these statistics, few of the world's 1.5 billion English speakers are fluent. Most get by with a vocabulary of a couple thousand words, as compared to the eighty thousand familiar to the average American or Briton. Pronunciations are often simplified, especially in the case of tricky consonant clusters. (For example, *cluster* becomes *clusser*.) Rules of grammar are frequently streamlined, irregularities dropped. (For instance, the *-s* at the end of third-person singular verbs is omitted: *I omit, you omit, he omit.*) And the lexicon is often supplemented with imported words, such as *sudoku* from Japanese and *hohlraum* from German, or hybrids such as *burquini* combining Arabic and English.

Observing these adjustments to the language and noting the vast number of English speakers making them, many linguists believe they're harbingers of how English will be spoken in the future. Retired San Diego State University professor Suzette Haden Elgin has dubbed this language-by-consensus *Panglish*, a term she used allusively in a couple of science fiction stories before dropping it into an interview for a 2008 *New Scientist* article on the evolution of English: "I don't see any way we can know whether the ultimate result of what's going on now will be Panglish—a single English that would have dialects but would display at least a rough consensus about its grammar—or scores of wildly varying Englishes all around the globe, many or most of them heading toward mutual unintelligibility." Dampened with academic caution, her comment was buried in the article's thirty-second paragraph.

But newspapers seldom miss a divisive term. Two days before the issue of *New Scientist* was officially published, *Panglish* was in the headlines. "English Will Turn into Panglish in 100 Years," announced the *Telegraph*; "Soon We Will All Speak Panglish," proclaimed the *Sun*; "How English As We Know It Is Disappearing . . . to Be Replaced by 'Panglish,'" declared the *Daily Mail*. Anyone who missed the headline sensationalism could find more of the same in the articles. "It is English but not as we know it," wrote the *Daily Mail* science correspondent David Derbyshire, for example. "A new global tongue called 'Panglish' is expected to take over in the decades ahead, experts say. Linguists say the language of Shakespeare and Dickens is evolving into a new, simplified form of English which will be spoken by billions of people around the world." Readers took the bait and expressed their outrage. Amid complaints about pinched vowels and nouns treated as verbs were broader statements of intolerance. "The problem is that people teaching English may not be English themselves," commented Jan from London, citing Irishmen and New Zealanders as interlopers. "Language teaching is big business and the rest of the world seems to be making money out of our language." Alastair from Corvallis, Oregon, was more succinct: "Panglish equals ignorance."

Of course linguistic nativism, like its political counterpart, is not a new phenomenon. Sir John Cheke's polemic was one of many in the sixteenth-century "Inkhorn Controversy," so named because Cheke and his compatriots deemed words of continental origin to be obscure, as "darke" as a hornful of ink. This chauvinist impulse is perhaps harshest in Sir George

Gascoigne's assertion, "The most auncient English wordes are of one sillable, so that the more monasyllables that you use, the truer Englishman you shall seeme, and the less you shall smell of the Inkehorne."* In spite of his French name, Gascoigne was still spoiling in 1575 to avenge William the Conqueror's eleventh-century invasion.

But if the Panglish panic is a tabloid reprise of the Inkhorn Controversy on the nativist side, the debate differs markedly on the part of those who advocate this latest stage of evolution, since the vast majority don't come from English-speaking nations. One of the strongest assertions comes from the Frenchman Jean Paul Nerrière. "Anglophones no longer own English," he told the London *Times* in 2006. "It is now owned by people in Singapore, Ulan Bator, Montevideo, Beijing and elsewhere."

"Elsewhere" in Nerrière's case is Provence, from which place he staked his claim to future English in 2004 by trimming the grammar to six verb tenses and the vocabulary to fifteen hundred words, while also eliminating idioms, metaphors, and humor.† He dubbed this low-impact English *Globish* and having trademarked the name started marketing his system as "the worldwide dialect of the third millennium" through books and computer software directed to fellow Europeans, Asians, and Indians.

His system has some foundation. As an IBM international marketing executive in the 1980s, Nerrière observed that he

---

* The principles of Gascoigne and Cheke have recently been taken up in a lighter vein on the Anglish Moot, a wiki devoted to exploring "English without words borrowed from other languages," which has posted "Anglish" versions of Hamlet's soliloquy and, somewhat more perversely, Martin Luther King's "I Have a Dream" speech, the latter rendered in part, "I have a mindsight that one day Mississippi shire, a shire sweltering under downtrodden-ness' heat, will be shaped otherwisely into an lush well, brimming with freedom and fairness."

† The lexical and grammatical limitations are negotiated by paraphrasing. Siblings are referred to as "the other children of my mother and father," *kitchen* becomes "the room where you cook food," and a mouse is called "the animal chased by cats." Given these circumlocutions, idioms, metaphors, and humor don't stand a chance.

was able to communicate more effectively with other non-Anglophones than native English speakers could communicate with them.The reason, he believed, was that the language of native English speakers was often unintentionally complex, and they didn't know how to simplify it, whereas most non-Anglophones had independently stumbled on a simplified version comprehensible to everyone.The origin of Globish was Nerrière's attempt to codify this vernacular, which he argued would sharpen its effectiveness as a tool, since everyone would share the same pool of words and rules. He envisioned Globish being taught with a brief 182 hours of instruction, and even foresaw Anglophones learning it in order to make themselves understood worldwide.

Nerrière has been adamant in interviews that Globish "is not a language" because a language "carries a heritage coming from history" and "is a vehicle for culture," whereas Globish is "a proletarian and popular idiom" necessarily limited, merely bringing people to "the threshold of understanding." He has also insisted that such limitations help to preserve the mother tongues of Globish speakers; finding Globish inadequate for deep personal expression, people will associate their culture and heritage and history with French or Dutch or Japanese.

Globish has done well enough as a business, as might be expected of any enterprise promising a basic grasp of English in 182 hours. As "the worldwide dialect of the third millennium," though, Globish has had no discernable impact.* The reason is

---

* The failure of Globish should come as no surprise, given the fate of a conceptual predecessor, Basic English. Formulated in the late 1920s by the eminent British scholars C. K. Ogden and I. A. Richards, the 850-word "International Auxiliary Language . . . for all who do not already speak English" was vigorously endorsed by Winston Churchill. In spite of this Ogden's ultimate goal that Basic become "the Universal language of the world" collapsed with the British Empire.While aspects of Basic informed the Special English of Voice of America radio broadcasts, the greatest impact may have been negative, encouraging George Orwell to lambaste linguistic universalism with his satirical Newspeak.

that, for all his keen observation of global speech patterns, Nerrière has ignored their basis and the broader historical context of these changes. He is right that "Anglophones no longer own English" but wrong to believe that they ever did, or that, trademark notwithstanding, he can take possession of it. Globish may bear a superficial resemblance to Panglish, but fundamentally they're opposites. With a fixed vocabulary centrally regulated like French, Globish is proprietary, whereas English is ubiquitous and open-source. This is how it became Panglish, or rather why it has always been Panglish.

It's also why English is so frightening to those intent on preserving other languages. Unlike the closed system of Globish, Panglish cannot be confined. The ease with which English absorbs words and even grammar from foreign tongues—such as the assimilation of Spanish through the Spanglish spoken in Harlem and Los Angeles—ensures that those attempting to banish it are facing a new permutation every moment. All other languages are to be found within English. Any serious attempt to "purify" Spanish or French of English influences is likely to result in an identity crisis. Moreover even if their respective academies ban English words from their respective dictionaries, the guardians of other languages cannot make their dictionaries off-limits to English. Nor can they restrict new meanings brought back when Anglicized vocabularies repatriate. In a globalized world of Panglish there may be linguistic diversity, but no longer is there linguistic particularity.

Thus the future of English is the future of language. This is not a matter of the English lexicon or rules of grammar obliterating others, as is so broadly feared. Rather the engine of English, which roughly speaking is the mechanism of entropy, is destined to be—in a world as interconnected as ours has become—the mechanism of language generally. Academies are anachronistic and obsolete.

Nearly five centuries ago Sir John Cheke made the mistake, repeated by the guardians of foreign languages today, of equating linguistic borrowing with bankruptcy. As the history of English since then has demonstrated, language does not follow the rules of accounting. A better analogy is macroeconomics, in which debt is productive, though even this comparison is imperfect, since in language supply and demand are mutually reinforcing. Language is richest when it is freely circulated.

Index